Khaoula Abrougui

Mechanization: accurate information on agricultural equipment

Khaoula Abrougui

Mechanization: accurate information on agricultural equipment

ScienciaScripts

Imprint

Any brand names and product names mentioned in this book are subject to trademark, brand or patent protection and are trademarks or registered trademarks of their respective holders. The use of brand names, product names, common names, trade names, product descriptions etc. even without a particular marking in this work is in no way to be construed to mean that such names may be regarded as unrestricted in respect of trademark and brand protection legislation and could thus be used by anyone.

Cover image: www.ingimage.com

This book is a translation from the original published under ISBN 978-620-2-27851-5.

Publisher:
Sciencia Scripts
is a trademark of
Dodo Books Indian Ocean Ltd. and OmniScriptum S.R.L publishing group

120 High Road, East Finchley, London, N2 9ED, United Kingdom
Str. Armeneasca 28/1, office 1, Chisinau MD-2012, Republic of Moldova, Europe
Printed at: see last page
ISBN: 978-620-5-92121-0

Table of contents :

General introduction

Mechanization is certainly one of the most important keys to the agricultural economy, and today it has become widespread in all sectors of agriculture. The constant and rapid evolution of agricultural equipment offers the farmer numerous options among which he must make choices that will condition his technical and economic performance.

- year 1900 : use of iron in the construction of agricultural machines with animal traction (ploughs, harrows...);

- 1920-1930: appearance of the first tractors with 2-stroke engines which replaced horses in front of the dragged tools;

- Late 1930s: Mr. Harry Ferguson built and marketed the first tractors with 4-stroke engines and 3-point hydraulic lift, following the development of mounted implements;

- year 1960 : appearance of the 4 wheel drive tractor ;

- 1980-1990: the improvement of agricultural machinery continues in several major directions (development of hydraulics, transmissions and electronics);

- Today, these improvements continue in several directions (automation, tractor-implement linkage, data recording).

All of these developments help to improve the performance of agricultural machinery and make it easier to drive. It is therefore essential to have accurate and nuanced information on agricultural equipment.

Chapter 1: Engines and Tractors

The role of an engine is to transform the heat energy contained in a fuel into mechanical energy (rotation of a shaft: crankshaft). It is a set of mechanisms that serves to transform energy.

1. components of a motor (Fig 1, 2)

1.1. Fixed bodies

- **Engine block**

It is the central part of the engine, also known as the cylinder block, which contains the other parts of the engine (cylinder, piston, crankshaft, camshaft, etc.) and has galleries where the engine coolant and oil circulate (Fig. 3).

- **Cylinder head**

It closes the engine block or cylinder block at its upper part and in which are installed the valves, the rocker arms, the rocker arm rods, the coolant passages, the injectors (MD), the fixing screws on the block, the combustion chambers and the oil passages (lubrication of the rocker arms). A cylinder head gasket is installed to ensure tightness and prevent leaks.

- **Carter**

A rocker arm cover or upper casing hides the cylinder head to avoid its exposure to the atmosphere and humidity. It contains an oil filler cap and a breather or tube for the release of gases (air).

The lower housing or oil reservoir closes the engine at the bottom and contains the oil necessary for lubrication. A crankcase seal prevents any oil leakage (Fig 4)

NB: each part has an optimal temperature to resist. Gaskets can withstand high temperatures up to a certain limit, beyond which the gasket melts. In this case, we change the gasket and not the cylinder head or the block. The gasket is designed to prevent leaks, withstand high temperatures and protect the engine components. The rise in T is due to the absence of cooling water.

1.2. Moving parts

- **Piston-cone** (Fig 5)

Pistons are cylinders on which the thrust of the explosion is exerted. They can be mounted in several ways in the engine block:

- Bored block (without cylinder, no risk of coolant leakage but change of engine block in case of wear) ;
- Dry liner (replace the liner if it is worn and therefore keep the engine block);
- Wet liner (good cooling but good sealing is needed to prevent leakage) (Fig 6, 7)

Bored blocks are becoming more common because of their lower cost. The efficiency of each assembly depends mainly on the materials used and the quality of the assembly.

The seal between the piston and the liner is ensured by three rings: a fire ring, a sealing ring and an oil scraper ring (Fig. 8)

The connecting rods connect the piston to the crankshaft and thus convert the reciprocating motion of the piston into rotary motion. The connecting rod foot is connected to the piston and the connecting rod head to the crankshaft

(Fig 9)

- **Crankshaft**

The bearings, called journals, are located on the axis of rotation and the crankpins on which the connecting rods are fixed. To reduce friction, the crankpins and journals are mounted on two half-bearings. The shape of the crankshaft depends on the firing order of the engine and does not tolerate any deformation, otherwise it must be ground or changed. The machining precision is in the order of a hundredth of a millimeter (Fig 10).

- **Flywheel**

To reduce vibrations as much as possible, manufacturers attach a metal disc called a flywheel to the crankshaft at the engine's output to regulate speed.

- **Camshaft**

It is located on the side of the engine at mid-height (side camshaft).

It is directly installed in a bore of the block. It is equipped with pushrods, rocker rods and rocker arms that control the opening and closing of the valves.

Tractor engines have two valves per cylinder (one intake and one exhaust), so there must be two cams on the shaft per engine cylinder. It usually drives other equipment (diesel pump, engine oil pump...) (Fig 11)

NB : The distribution organs (camshaft, push rods, rocker arms, valves)

2. Engine operation

2.1. Motor characteristics

- TDC is the highest position of the piston.
- The stroke is the distance between the TDC and the TDC, it is also the length of piston travel or the diameter of the crankshaft.
- The bore is the inside diameter of the cylinder.

- The unit displacement (V) is the volume of the cylinder between TDC and BDC.
- The combustion chamber (v) is the volume of the cylinder located above TDC (**Fig 12**)
- The engine speed (rpm) indicates the speed of the crankshaft (700-2200 rpm).
- The tractor engines have 3 to 6 cylinders in line depending on their power:
- 3 cylinders : 40-60 hp 4 cylinders : 60-90 hp 5 cylinders : 90 hp 6 cylinders : 100 hp and more.

2.2. Operating cycle of a 4-stroke engine (diesel engine)

The operation takes place on a cycle of 4 successive phases or 4 times.

■ **Admission**

During this phase the piston moves from TDC to BDC, this descent is followed by an increase in volume and therefore a depression, the intake valve opens to let the filtered air from outside enter. When the piston reaches TDC, the valve closes.

■ **Compression**

Both valves are closed. The piston moves up from TDC to TDC, the volume decreases so the pressure increases in the combustion chamber. The T of the air increases because of the high compression ratio.

■ **Explosion-relaxation**

When the piston is at TDC, an injector sprays the hot, compressed air with fine droplets of diesel. Combustion starts without a spark. The very high pressure pushes the piston towards TDC. Both valves are closed.

■ **Exhaust**

The exhaust valve opens when the piston is at TDC. The piston moves up to discharge the exhaust or burnt gases to the outside. The exhaust valve closes at TDC and a new cycle can begin (**Fig 13**)

NB: an engine stroke corresponds to a rise or a fall of the piston (one engine stroke = one stroke = ^ crankshaft revolution). A four-stroke cycle therefore takes place over 2 crankshaft revolutions and one camshaft revolution since each valve opens only once.

The camshaft is said to rotate at half speed of the crankshaft. If an engine is running at a speed of 2000 rpm, 1000 explosions per minute per cylinder occur.

2.3. Operating cycle of a 4-stroke engine (gasoline engine)

■ During the intake phase, the depression created in the cylinder following the descent of the piston from TDC to BDC and the opening of the intake valve (exhaust valve closed), leads to the penetration of a dosed mixture (air + gasoline) into the cylinder.

- During the compression phase, both valves are closed, the piston rises from TDC to TDC, the volume decreases, the pressure increases but the T inside the cylinder remains relatively low to make the ignition. Hence the need to create a spark to ignite the quantity of gasoline. This spark is obtained by a spark plug. Gasoline engines are called "spark ignition engines".

- The piston is moved during the explosion-expansion phase from TDC to BDC, this phase corresponds to the engine time that activates the piston following the pressure generated by the ignition of the gasoline. Both valves are closed.

- The exhaust valve opens during the exhaust phase, the burnt gases are discharged to the outside and the cycle starts again.

NB: a 4-stroke engine needs 4 phases to work, a 2-stroke engine needs two phases to work (the 4 phases are two by two), i.e. one engine time corresponds to 1 crankshaft revolution and not 1/2 revolution, so it is faster than a 4-stroke engine.

2.4. Ignition order of a 4-stroke and 4-cylinder engine

Cylinders	1	2	3	4
^ t (vilebr)	**D** (exp-det)	C (comp)	E(escape)	A (adm)
^	E	**D**	A	C
^	A	E	C	**D**
^	C	A	**D**	E

The order of ignition (on explosion) is: 1, 2, 4, 3

2.5. Distribution

Timing is the drive of the camshaft from the crankshaft. On diesel engines used in agriculture it is provided by three pinions (**Fig 14**). Two other types of distribution exist on motor vehicles and not on tractors (by chain or by belt that can be changed every 60,000 to 80,000 km).

3. Circuits necessary for the operation of a motor

3.1. Specific circuits for gasoline engines

- **Fuel supply system**

A carburettor is a device used in gasoline engines, its role is to dose the gasoline and the air that feeds the cylinder when the piston slides from TDC to BDC during the intake phase following the depression created in the cylinder. In a carburetor there are 4 essential circuits:

Cold start circuit (more gasoline), normal running circuit (1 g of gasoline for 18 g of air), recovery circuit (rich in gasoline), idle circuit (lean mixture in gasoline) (Fig 15)

■ Ignition circuit

It is formed by the following essential components: Battery (current accumulator), coil, high voltage current distributor to the spark plugs according to the ignition order, spark plugs (Fig 16).

3.2. Specific circuits for diesel engines

■ **Diesel supply circuit (one low pressure circuit and one high pressure circuit)** (Fig 17, 18)

The high-pressure circuit includes an injection pump supplied with diesel by the low-pressure circuit. Its role is to dose the quantity of diesel necessary for an explosion and then to distribute the doses to the cylinders, at the right time and in the right ignition order. The high pressure of the diesel at the pump outlet is due to the injectors.

■ **Compression engine** (Fig 19)

The turbocharger increases the power delivered by an engine by improving the air filling of the cylinder. It can increase power by about 20 to 40% on a tractor diesel engine and 40 to 80% on a car engine. The turbine is driven by the exhaust gas pressure (80,000 to 100,000 rpm) and the compressor by the turbine.

■ **Preheating circuit**

It is formed by resistors. These are current-fed heating plugs. There can be one spark plug per cylinder or one to heat the air at the inlet of the intake manifold **Fig 20.**

NB: combustion takes place without sparks because the diesel is sprayed into sufficiently compressed and hot air. This is called auto-ignition by compression.

3.3. Common circuits

■ **Lubrication circuit Fig 21, 22**

The oil stored in the lower crankcase has several roles:

- o Lubricate moving parts to avoid metal to metal contact
- o Cooling moving parts
- o Clean the engine of all impurities from the combustion process.

The oil pump in the crankcase is usually located at the front because it is driven by the timing system.

■ **Cooling system Fig 23**

Water cooling :

Water circulates in the engine at the cylinders and cylinder head. The water pump (centrifugal pump) is driven by a belt from the crankshaft. It runs continuously when the engine is running.

The thermostat is a thermostatic valve that opens when the engine water is hot and needs to go to the radiator to cool down. It closes if the T of the liquid is lower than the normal T of the engine. It allows to maintain a constant T in the engine of about 90°C.

The fan is driven by a belt from the crankshaft.

Air cooling: Fig 24

This type of cooling uses an external air circulation with a lower conductivity than water. The exchange surfaces must be increased (fins) and the air flow must be very high. Disadvantages:

- o Noisy turbine (driven by a belt from the crankshaft)
- o Engine T less well stabilized than in the case of water cooling Engine T variations modifying the fuel combustion conditions (more polluting exhaust gases).

- ■ **Starting and air filtration circuit**

4. Performance of an engine

The performance of an engine is presented in the form of curves because they vary according to the engine speed **(Fig 25)**.

Torque is the force transmitted in rotation from the crankshaft to the wheels and the power take-off. It results from the force of the gas pushing on the piston during the explosion. It varies with engine speed because it depends on the filling of the cylinders and the quality of the air-fuel mixture. Starting from idle speed, the torque increases to a maximum value.

The torque is measured with a braking device that is connected to the flywheel or to the power take-off (on tractors).

The power (watt) is equal to the product of the motor torque (mxN) and the rotation speed (radians/sec).

The horse is a common unit **1 horse = 735.35 W**

The power varies (increases) with the engine speed.

5. Power transmissions on wheeled agricultural tractors

The tractor's transmission is used to transmit or interrupt engine power to the wheels and power take-offs (PTOs).

Starting from the engine, we find in order: the clutch, the gearbox and the rear axle (a bevel gear, a differential, wheel shafts, reducers and brakes) **Fig 26, 27**.

5.1. Clutch

It is a mechanism that allows the transmission of the rotational movement of the crankshaft to the input shaft (primary shaft) of the gearbox and the rest of the transmission or to cut the movement. It consists of a disc, a clamping plate and a stop **(Fig 28, 29)**. The clutch control consists of a pedal, rod, connecting rod, bearing and a diaphragm spring in the clutch plate. Most clutches are of the dry single-plate type (**Fig 30**).

At rest, the springs extend pushing the plate forward and then the disc against the flywheel making them interdependent (the disc is clamped between the clamping plate and the flywheel, the movement is transmitted to the gearbox: clutch position). To stop the transmission of the movement, you must oppose the springs by compressing them. By pressing the clutch pedal, the stop moves, presses on the fingers which, by pivoting, bring the pressure plate back (the disc is free, no movement is transmitted: disengaged position). Diaphragm clutches **(Fig. 29)** simply have a diaphragm (elastic disc with many fingers) instead of the 3 fingers **(Fig. 28).**

The tractors have a dual clutch **(Fig 31)** comprising:

- A main clutch for the advancement ;

- A clutch for the power take-off.

When the clutch pedal is pressed halfway down, both pressure plates 7 and 8 move to the right, thus releasing clutch disc No. 9 from the power take-off. When the pedal is fully depressed, only pressure plate 7 moves to the right (pressure plate 8 is stopped by the stop 13). The drive plate is released and the vehicle stops.

5.2. Gearbox

It allows to modify the ratio between the rotation speed of the crankshaft and that of its output shaft (which will be transmitted to the wheels). It allows to decrease the speed but amplifies the CM engine torque ; It is a torque transformer.

The gearbox is formed by three shafts (primary, secondary or output and intermediate).

The primary shaft starts from the clutch disc to the first small free gear of the gearbox. The gears of the primary and intermediate shafts are always in mesh (geared), which causes the primary to drive the intermediate **(Fig. 32)**.

The wheels that rotate freely in the output shaft are called idler wheels and the wheels that rotate freely and translate freely are called sliding sleeves. To transmit the motion to the secondary shaft, the impeller must be associated with the sleeve.

In the BV there are 4 forward gears and one reverse gear.

Exercise:

4 motor revolutions1 .... revolution at the output Cs = 4

Cm **1stR1** .... = 0,25

2.5 motor revolutions1   revolution at the outputCs   = 2.5 Cm   **2 ndR2**

.... = 0,40

1,5 tr motor1 .... turn at outputCs   = 1,5 Cm   **3rdR3** ...... = 0,66

1 tr motor .... 1 turn at the output (direct drive without passing through the shaft

$$Cs = Cm \ \textbf{4th R4} = 1$$

$\textbf{P} = \textbf{C x W}$ = constant. P : power, C : force torque, W : speed of rotation

$P_M = PS$ (Motor power or input power = output power)

Cm x Wm = Cs x Ws $\textbf{Cs = (Wm/ Ws) x Cm}$

The ratio R = ...Cs/Cm4   front reduction ratio.

o The reverse gear and the first gear have the same ratio.

3.4. Differential

The differential role is to allow a driving wheel to turn faster than the other wheel on the same axle. This is essential when cornering (the outside wheel turns faster).

Each of the wheels is connected to a half-axle with a pinion at its other end, called a sun gear. The two sun gears housed in the differential case are connected to each other by two or more planet gears **(Fig 34, 35)**.

The large ring gear of the bevel gear is also fixed to the housing. For straight line travel, the gearbox rotates with the ring gear, the two planet gears are fixed in rotation. The two planetary gears rotate at the speed of the ring gear, i.e. the speed of the wheels. On the other hand, in a curve, a wheel can turn faster because the satellites turn on their axes.

The main disadvantage of the differential is when the two wheels are not on the same ground (one wheel at the bottom of the furrow when plowing and the other wheel on the ford). The wheel at the bottom of the furrow offers a high resistance. At this moment the satellite turns on its axis and around the immobilized sun gear. All the movement is thus transmitted to the wheel with the least resistance (the one on the ford) and the tractor skids on the spot. It is then necessary to use the differential lock mechanism that allows the two wheel halves to be connected (the two wheels turn at the same speed) **Fig 36**. This mechanical system must be engaged at a standstill or at low speed. It disengages automatically as soon as the pedal is released.

6. Agricultural tractors

- Main functions of a tractor

- **Front:** push, load, lift, animate
- **Rear:** pull, lift, animate

Tractors can be classified according to their use (multipurpose, horticultural, forestry), their size (small, medium and large tractors), their means of movement (tracked tractors, pneumatic tractors with 2 or 4 drive wheels) and their chassis (monobloc, articulated).

- **Advantages of tracked tractors**

- Power (tractive effort proportional to this mass) ;
- Greater adhesion (strength of grip on the ground);
- Good weight distribution over a large track area.

- **Disadvantages**

- No traffic on the road
- Low speed of movement

- **Components**

- Engine
- Frame (sum of parts) with rigid or articulated mounting
- Tires
- Means of control
- Transmission
- comfort

## 6.1.	Track adjustment of an agricultural tractor

The track of a tractor is the distance between the center planes of the two wheels on the same axle. The track of a tractor can be adjusted in order to conduct maintenance operations on row crops.

- By translation of the wheel hub on its axis
- By turning the rim and/or the sail with asymmetrical cleats
- By screwing the rim to the rim (screw (rim) - nut (rim) system)

## 6.2.	Improving the grip of wheeled agricultural tractors

A wheel that sticks to the ground moves forward by the same amount as it turns (the equivalent of its perimeter for one revolution), but a wheel that does not stick to the ground moves forward by less than it turns. We must therefore improve the grip of a tractor to increase its pulling force.

■ **By weight increase**

- By adding weights to the wheels
- By inflating the driving wheels with water (200 l/tire max)
- By transferring the load from the tool to the tractor

■ **By increasing the wheel-ground contact area**

- reducing wheel inflation pressure to acceptable limits
- by pairing of driving wheels
- by locking the differential (to turn both wheels)

■ **By changing the nature of the wheel material**

- By wrapping the drive wheels with a chain

6.3.   Hydraulic lift

The hydraulic lift is a 3-point type. It consists of two lower lifting arms and an upper push bar, often called the third point.

$$F = Rs \times S$$

$F = Rs \times L \times P$

Rs = specific soil resistance, S = working section, L = working width, P = ploughing depth

Hitching mode

- **3-point linkage**: mounted implement for transport (raised implement for transport)
- no wheels or skids: carried to work (all weight on the tractor)
- wheels or skates: semi-mounted at work
- **1 single hitch point** (towed implement): 4-wheel trailer (all weight rests on the ground)
- Two wheels: semi-mounted for transport (the front part rests on the tractor and the rear part rests on the ground.

The hydraulic lift only concerns three-point hitched implements, i.e. implements mounted for transport and mounted or semi-mounted for work.

- **Position control**

This type of control is used for tools carried for transport and work and keeping a fixed distance from the ground (fertilizer spreader, sprayer);

- **Effort control**

It is used for tools that work in the soil and have no means of support (wheels or sliding blocks (ploughs with shares and mouldboards)

- **Mixed control**

It is the combination of the two controls effort and position. It is also used for tools that work in the soil and have no means of support, in case of an obstacle the effort control is added to escape this obstacle and not to break the tool (plough with shares and mouldboards).

- ❖ **Floating position**

It is used for tools that work in the soil (seeder, rotavator) or on the surface of the soil (mower) and that have means of support on the ground (wheels or skids).

In the floating position, the tractor-implement linkage is an independent linkage, i.e. if the tractor floats to the right or left because of an obstacle, the seed drill is not affected and continues to do its work normally to better bury the seeds in the soil.

6.4.   Motion capture

- Towed machine
- Wheel driven machine
- Machine driven by carrying and driving wheels
- Machine driven by PTO or PDf/PTO
- Self-propelled machine
- Machine driven by pressure wheels

Chapter 2: Study of the tools for setting up the crops

Introduction

Mechanization is certainly one of the most important keys to the agricultural economy, especially since the profitability of our different speculations is seriously threatened. The constant and rapid evolution of agricultural equipment presents the farmer with numerous options among which he must make choices that will condition his technical and economic performance. It is therefore essential to have accurate and detailed information on agricultural equipment.

I. Ploughing tools (Ploughs)

1. Ploughs with shares and mouldboards

Plowing is an operation of turning over the soil to an average depth of about 20 cm. Its purpose is to create a structural state of the soil favorable to the development of plants by facilitating germination, growth and installation of roots. It has several objectives:

- Soil loosening and aeration;

- landfill of organic materials;

- destruction of weeds.

Ploughing is therefore a fundamental soil working operation carried out in the vast majority of cases with a ploughshare before the planting of a crop. Theoretically, it consists of cutting a strip of land with a rectangular section and turning it over so as to bring the lower part to the surface **(Fig 1)**.

The vertical part that remains after the passage of the plow is called "wall", the horizontal part is called "bottom of furrow" or "gauge" and the part of the field not plowed is designated as "ford" **(Fig 1)**. The ploughing depth varies from 15 to 30 cm depending on the type of soil, the date of ploughing, the farmer's cultivation techniques... The width of a ploughing strip is measured in inches (1 inch = 2.54 cm). The most commonly used plows have share widths (or furrow widths) of 12, 14, 16 and 18 inches (30 to 45 cm). The width is usually fixed for a plow except for the so-called "vari-width" plows where the width is hydraulically variable for most.

By changing the depth, the farmer can change the appearance of the plowing done:

Upright ploughing: this is a deep ploughing that will produce square shaped strips. The aeration is important while the turning is insufficient. This ploughing is often used as winter ploughing to facilitate the infiltration of rainwater.

$P/L = 1$

Ploughing: a shallow ploughing that will produce very rectangular strips. Turning is important, facilitating soil preparation (**Fig 2**).

P/L = 0.5

Flat ploughing: a correct ploughing will give a good turning over and a good aeration.

P/L = 0.75

As far as plowing techniques are concerned, a distinction must be made between flat plowing and plank plowing **(Fig. 3)**. Flat ploughing is obtained with reversible ploughs that can alternately plough to the right and to the left, so as to lay down all the cut strips of a field on the same side. They leave a ground without ados, nor derailment. Ploughing is done with ploughs that can only pour the soil on one side. They are carried out either by starting from the edges of a field to plough by returning towards the center, these are the ploughings known as "en refendant", or by starting from the middle of this field to plough by turning around this median line, they are then ploughings known as "en adossant".

1.1. Main parts of the plough

A plough with shares and mouldboards has one or more plough bodies rigidly fixed by means of connecting parts (age, strut) on a frame connected to the tractor by a headstock which allows its rotation in the case of reversible ploughs.

The plough body is the main element of the plough, it includes the working parts (share, mouldboard, skimmer, coulter) and the protection parts (sep, counter-sep, heel) and often an adjustable reinforcement ensuring the rigidity of the whole **(Fig 4)**.

1.1.1. The working parts a- The share

The function of the share is to cut horizontally the strip of land to be turned by the mouldboard. It is usually trapezoidal in shape, which lends itself best to a suitable connection with the mouldboard. Depending on the cross-section for the front side wing, the following categories are distinguished **(Fig. 5):**

■ straight cutting shares with reinforced point without particular shape, reserved for not very abrasive grounds;

■ Blade plowshares, sometimes reversible, mainly used in soils without penetration difficulties;

■ duckbill coulters with a flared tip, particularly suitable for difficult and abrasive soils;

■ the movable point shares with a steel bar of square section which engages in a housing arranged at the front of the share; they are suitable for abrasive and difficult to penetrate soils;

■ shares with removable tip, whose tip is reversible in case of wear;

■ The two-piece coulters with a special steel tip attached to an ordinary steel tip are designed to work in all types of soil.

b- The moldboard

The function of the mouldboard is to lift and turn the soil strip. In order to be light and flexible, strong and resistant to wear, mouldboards are most often made of triplex steel, which is a metal formed of three layers. The mode of action of the mouldboard depends on its shape and certain characteristic angles. According to their geometrical shape, moldboards can be classified into three main families **(Fig 6).**

- helical mouldboards are desired for winter ploughing giving a cloddy and slightly dyked ploughing. They are especially suitable for sticky soils;
- the cylindrical mouldboards give an important loosening favorable to spring ploughing when they are long. When they are short and flat, they give an upright ploughing recommended for winter ploughing;
- Cylinder-helical mouldboards, often called "universal" or "American", have a cylindrical front part and a helical rear part. They generally give a ploughing with a cloddy base and a crumbled surface.
- the slatted mouldboards are made up of independent and interchangeable blades. They are recommended in greasy and sticky soils, where they facilitate the take-off of the soil strip and require a slightly lower traction effort than conventional mouldboards.

c- The mouldboard extension

The mouldboard tail attached to the rear part of the short mouldboard increases the turning effect and improves the parting clearance (**Fig 7**).

d- The coulter

The coulter is intended to cut vertically the strip of land that will have to be turned over and to make it free on the side of the ford, it thus traces the wall. There are three types of cutters (**Fig 8**):

- the fixed straight coulter is less and less used because it increases considerably the traction resistance, especially in dry soil;
- the mobile circular cutter is a flat disc with a smooth or crenellated cutting edge not recommended for stony ground;
- the blade cutter (incorporated) is in order to reduce the effort of traction. It can be welded or bolted on the share or the counter-share or it can be incorporated in the mouldboard.

e- The shave

Generally called "peloir" or "avant-corps" is a ploughing body in model form composed of a share and a mouldboard. It is intended to bury plant debris and weeds so as to obtain, when the work is completed, a ploughing without any trace of vegetation on the surface of the soil. The skimmer is not an essential plough part, it is better to use it only for the last ploughing before sowing.

All these working parts form the ploughing body which is connected to the 'frame' or 'age' of the plough by a stanchion.

1.1.2.   Supporting parts

a-The age: it is the frame and the main supporting part of the plough and it is to it that the traction force is applied.

b-The stanchion: it is a support piece which serves to join the age to the main parts of the plough: the share and the mouldboard.

c-The sep: This part, fixed to the lower part of the stanchion, serves as a support for the ploughshare and mouldboard and helps to ensure the stability of the plough (**Fig 7**).

1.1.3.   The protection parts

a- The counter-sep: it is a horizontal plate which protects the sep on the side of the wall, against the wear by friction and allows the lateral guidance of the plough and the advancement of the plough body parallel to the axis of the tractor.

b-The heel: it is a removable part that is attached to the rear part of the sep and counter sep for the vertical guidance of the plough.

1.2. Classification of ploughs

The criteria generally used to classify the different plough models are the type of ploughing and the method of attachment.

1.2.1. Classification according to the type of ploughing

a-Simple ploughs

These ploughs have only one set of bodies that always pour the soil on the same side, usually to the right (**Fig 9**). To plow, it is necessary to turn around the plot to finish in the middle. This leaves a furrow in the middle of the plot delimiting two plots, hence the name ploughing in plots. When the plot is large, the farmer makes several planks. The following year, it will be necessary to plough in reverse order to start in the middle of the plot and finish on the edges. This type of ploughing is tedious to carry out in order to finish the plots well. It is mostly used in wet, undrained soils to eliminate excess water.

b- Reversible ploughs

A reversible plough has two rows of reversed shares arranged at 180 (most common) or 90° to each other (half turn and quarter turn). These ploughs allow the soil to be poured alternately on one side and then on the other, thus achieving a flat ploughing (without a furrow in the middle of the plot) (**Fig 9**). It is necessary to alternate the sides from one year to the next to avoid seeing the furrows several times in the same place.

1.2.2. Classification according to the mode of coupling

This classification is based on the way in which the connection with the tractor is designed and has evolved over time to meet the changes in agriculture.

a-The trailed ploughs: they were initially used with animal traction and then the first tractors. The frame is supported by wheels and pulled by the tractor.

b-Mounted ploughs: they are connected to the tractor by means of the three-point linkage. The digging up and the digging out are controlled by the tractor's hydraulic hitch. With this equipment, the number of bodies is generally limited by the power of the lift and by the danger of the tractor's pitching up. However, at work, they have the advantage of being able to transfer part of their weight onto the tractor and thus improve adherence (significant load transfer). To allow effective control of the working depth, models with three or more bodies generally have a gauge wheel at the rear (**Fig 10).**

c-Semi-mounted ploughs: these materials are hitched to the drawbars of the hitch and rest on the ground, both for transport and for work, by means of one or two carrying wheels placed in the middle or at the rear (**Fig. 11**). In spite of its technical disadvantage (less load transfer), this type of plough is in full development because it is very easy to use (quick coupling, easy adjustment and monitoring).

1.3. Settings

These adjustments concern the preparation of the tractor, the tractor-plough coupling and the various adjustments to be made before and during work.

1.3.1. Preparation of the tractor

a-Track of the tractor

The track of the tractor must be adapted to the type of plough, and more precisely, to its cutting width. When ploughing, it is the inner side of the tire that is in contact with the wall. The distance measured between the inner sides of the rear tires should not be confused with the tractor's track, which is the distance between the tire axes and corresponds in fact to this distance plus the width of the tire (**Fig. 12).** For ease of driving, the track of the front wheels must correspond to that of the rear wheels.

b-Hydraulic lifting

The weight of the plough must be adapted to the lifting capacity of the hydraulic system. To work safely, a lifting force of three times the weight of the plough should be available at the coupling points. The lower linkage arms should always be at the same height and length.

c-Tyres

For tires, it is essential that the inflation pressure is the same on both sides. On the other hand, it is necessary to know that the lowering of the pressure allows a better contact of the tire with the ground. It is therefore in your interest to adopt a "ploughing" pressure, generally between 0.6 and 1.2 bars. These values vary according to the type and size of the tires used and it is necessary to refer to the indications in the tractor's maintenance booklet.

d-Weighting masses

Sometimes it is necessary to weight the rear of the tractor to increase its grip. In this case, it is possible to use ballast weights or tire ballast.

1.3.2. Hitching the plough to the tractor

On most tractors and ploughs, the attachment points of the top and bottom links can be moved, so these points should be chosen to ensure good stability of the whole linkage and optimum grip of the tractor.

1.3.3. Operating settings

The different adjustments to be made in the field are interdependent and will be analyzed in the order of execution.

a. Adjustment of the working depth

For all mounted ploughs, the ploughing depth is obtained by the height position of the lifting arms controlled by the tractor hydraulics, and is completed by a gauge wheel mounted at the rear.

In order for all bodies to work at the same depth, the plough frame must be parallel to the ground. This is achieved by adjusting the length of the push bar (third point).

b. Adjusting the plumb line

The plumb line consists in having vertical stanchions (seen from the back of the plough), i.e. the plough works perpendicular to the ground. It is adjusted from the turnover stops on a reversible plough or with the length of the lifting arms for a simple plough.

c. Adjustment of the heel

The heeling consists in having vertical struts (seen on the side of the plough). It is obtained by lengthening or shortening the top bar of the ploughing attachment. Once these last two adjustments have been made, all the ploughshares must plough to the same depth.

d. Tilt and offset adjustment

To adjust the tip tilt, it is necessary to change the orientation of the frame in relation to the direction of travel (**Fig 13**). The age of the plough must be rotated in relation to the headstock. This adjustment makes it possible to obtain the alignment of the counter stock in relation to the wall.

The offset is the transverse displacement of the plough age in relation to the headstock. It is used to adjust the working width of the first share (**Fig 14**).

2. Disc ploughs

These materials are similar in design to the ploughs with ploughshares and mouldboards, but instead of the classic plough bodies, they use spherical cap-shaped discs with a diameter varying from 650 to 900 mm. The disc ploughs exist in simple form, pouring the soil on one side only, and in reversible version or with pivoting discs, pouring the soil alternatively on one side and on the other. The anti-drift action is ensured by a gauge wheel, equipped with a stabilizing disc that opposes the lateral thrusts of the tool (**Fig 15, 16**).

Despite all the innovations, the conventional plough remains a robust, simple, low-maintenance tool that is still widely used. Since the priority objectives of most farms are to save time and reduce costs, plowing is increasingly being questioned. The first step is to replace it by decompacting with a tine tool. This one does not reach the same objectives: indeed, it loosens the soil well but does not bury the organic matter and destroys less weeds. The second step is the suppression of ploughing to carry out a direct seeding without soil work or with a superficial work limited to the seeding strip.

II. tillage tools

According to their working depth, tillage tools can be classified into two categories:

- Pseudo-ploughing" tools: they perform a deep work without turning the soil, which precedes or even replaces ploughing;
- The so-called "shallow tillage" tools: they prepare the seedbed and their working depth is less than that of plowing.

We also distinguish between tools with teeth, discs or rollers, and tools driven by power take-off.

1. Tools with teeth

The decompaction tools are intended to break up and loosen the layers of soil that have been compacted at depth (ploughing sole, passage of heavy equipment), in order to restore optimal conditions for plant development. The proposed equipments are designated as subsoilers, decompactors or chisels; the working parts are teeth, blades or spades.

1.1. Subsoiler

It is composed of a frame with a three-point linkage on which a single tine is fixed. The clearance under the frame can exceed one meter. The working depth can reach 80 to 90 cm. The power required varies according to the type of soil but remains high, around 100 to 150 horsepower per tine. It is used to decompact the soil in depth. The tine is equipped with a share at its end and sometimes with wings in order to widen the decompacted area in a V shape **(Fig 17).**

1.2. Decompactor

It has a single row of tines arranged on a straight or V-shaped frame, thus allowing improved penetration into the soil as it is done progressively **(Fig 18)**. The working depth can reach 70 to 80 cm depending on the model. The objective is to decompact the soil and eliminate a compacted zone which frequently corresponds to the ploughing sole (located below the ploughing, between 20 and 40 cm deep). The power required is in the order of 30 to 40 horsepower per tine. The number of tines used varies according to the power available and the chosen speed of advance. The cracking and aeration of the soil varies according to many factors:

The type of soil and its humidity;

The shape of the tooth (straight or curved) and the share;

The presence of fins on the tooth;

The forward speed (about 3 to 5 km/h).

Many equipment are also available as options:

Gauge wheel for depth adjustment ;

Rear roller for surface leveling and depth adjustment;

Vertical opening disc positioned in front of each tooth.

1.3. Chisel

On this tool several rows of tines are fixed on the same frame. The clearance under the frame of about 40 to 50 cm allows to reach a working depth of about 30 cm. This allows a loosening of the soil to a depth similar to that of plowing. The chisel is used for stubble plowing because the tines remove weeds, for plowing resumption, or even to replace a plow (but without turning the soil). The power required can vary from 10 to 20 horsepower per tooth. Each tine consists of a stanchion supporting a share at its end. To meet all situations, manufacturers offer a wide range of equipment **(Fig 19)**.

1.4. Cultivator

Cultivators are tine tools with multiple functions: decompaction, extirpation, burying, plowing, loosening of deep or superficial layers (5 to 20 cm deep), preparation of a seedbed. Depending on the operations envisaged and the depth of work, a distinction is made between heavy cultivators or chisels and light cultivators, called vibrocultivators. The power required is low (3 to 5 horsepower per tine) and the spacing of the tines per row is adjustable according to the work required. Many choices are possible depending on the use and working conditions **(Fig 20, 21).**

Lightweight cultivators or vibratory cultivators are often equipped with single or double curved "S" shaped vibrating tines **(Fig 22).**

Self-propelled tine cultivators are designed for high-speed stubble cultivation and pre-seeding preparations. They are rotary cultivators with two horizontal rotors equipped with tines or blades that overlap and work in opposite directions, ensuring straw crushing, soil crumbling and good mixing of organic matter.

Rolling spade cultivators **(Fig 23)** have several trains of rolling spades arranged at an angle to the direction of travel. The spades are made up of sharp, curved blades, which are fitted in half on a shaft or bolted to a flange.

In addition, vibratory cultivators are often fitted with equipment to level and re-compact the soil, controlled by the tractor's PTO.

1.5. Oars

Harrows are very simple tools, essentially performing the following operations:

- loosen the soil superficially;
- clean the soil by bringing the weeds to the surface;
- Level the ground by cutting down ridges and furrows;
-fill seeds and fertilizers.

Depending on the shape and movement of the working parts, there are two main categories of harrows: tine harrows and rolling cage harrows.

The teeth are arranged in several rows (5 to 6) forming Z-shaped compartments. They are positioned in staggered rows in order to avoid blockages. The length of the teeth varies from 15 to 25 cm. The shape of the tines and the dimensions can be very variable **(Fig 24).**

Rolling cage harrows, also known as roto-harrows, have horizontal cages with perpendicular or oblique bars as working elements **(Fig. 25)** and are most often associated with tine cultivators, whose seedbed preparation action they complement.

In difficult conditions, the need for several passes at the same place has led to their replacement by PTO driven tools. Their main advantage is their low operating cost.

2. Rollers

They can be used for several different reasons:

- Soil compaction to reduce the porosity which can be an obstacle to the root development of crops;
- Soil compaction to promote the rise of water from the depth to the surface;
- crumbling of the soil by crushing of clods or aggregates;
- settling of the plants (cereals) for a better contact with the soil.

Thus, rollers come in a wide variety of types that differ in the shape, diameter, and weight of their working parts. They are used alone or in combination with other materials and are commonly assembled to form wide sets **(Fig 26).**

- **Smooth rollers or pressure rollers:** they consist of a series of cylindrical elements made of sheet metal or cast iron, rotating around an axis. They are mainly used to ensure a compacting of the surface of soils which do not fear the beating;

-Skeleton rollers: they are made up of narrow discs of the same diameter and leave a more undulating and cloddy soil than the smooth roller;

-Croskill" rollers: they consist of large and small diameter discs that alternate on the same axis. These very efficient rollers are recommended for cloddy soil.

-Spiral rollers: the functional part of these rollers is a flexible steel spiral bar mounted on a central axis. This type of roller is commonly associated with controlled soil cultivation tools mounted on the front of the tractor.

-Disc rollers: the principle and the use are similar to the previous type except that the spiral is replaced by large diameter discs assembled on the same axis.

-Star rollers: these rollers, similar to the previous ones, are made up of a variable number of star discs mounted on the same shaft **(Fig. 27).**

-Packer rollers: they are made up of a cylinder with radial teeth and associated with scraper blades. They are often integrated to the tools ordered.

-bar rollers: they comprise a series of rings supporting bars or tubes arranged along the generatrix or in a helix **(Fig 28).**

3.1. discs tils

Depending on the type of tillage required, there are three categories of disc tools:

- disc ploughs ;
- disc harrows ;

  - disc sprayers.

3.2. **Disc ploughs (Fig 15)**

3.3. **Disc harrows**

The disc harrow is a tool for shallow soil cultivation, it is mainly used for turning stubble after harvest. There are 9 to 14 discs and their diameter varies from 550 to 600 mm. To absorb the transverse thrust of the discs, the tool is equipped at the rear with one or two stabilizing wheels **(Fig. 29).**

3.4. **Disc sprayers**

Disc harrows are multipurpose tools; they are used for stubble cultivation, ploughing (they cut strips of soil and then partially turn them over by throwing them to one side) and seedbed preparation.

Depending on the size of the discs, the arrangement and the number of disc trains, a distinction is made:

3.5. Offset sprayers **(Fig 30)**

3.6. Cover-crop" sprayers

3.7. Tandem" sprayers **(Fig 31)**

The dimensions of the discs allow to classify the sprayers in three ranges:

Range	Weight/disc	Diameter (cm)	Thickness (mm)	Spacing (cm)
Slight	<60 kg	46 à 51	3 à 4	16 à 23
Average	60 to 80 kg	56 à 61	5 à 6	21 à 23
Heavy	> 80 kg	66 and over	5 à 10	23 à 26

4. PTO driven tools

The tools whose working parts are driven by the tractor's power take-off are designed to carry out the so-called pseudo ploughing work, in preparation of the germination beds.

Compared to conventional pronged instruments, animated tools have the following advantages:

- a good adaptation to compact and difficult to work soils;

- better use of the tractor's power;

- the possibility of association with other materials, to carry out in one passage, the resumption of ploughing, the preparation of the germination bed and the sowing.

A classification based on the type and mode of action of the working organs, allows to distinguish :

- **Rotary cultivators**: this tillage tool is commonly called a Rotavator **(Fig 32)**. Several settings can be used to determine the degree of crumbling or fineness of work:

- the speed of advance,
- the rotor speed,
- the number of blades per crown,
- the position of the rear apron (folding),

For depth adjustment, all machines are equipped with skids or gauge wheels.

- **rotary harrows (Fig. 33):** this is the most common tool with tines driven by a power take-off on farms. Their great interest is the recovery in a single pass of clay soils with low humidity, without deep compaction. More and more often, these machines are combined with a seeder in order to carry out the preparation and seeding in a single pass. With all rotary harrows, the working depth is adjusted by means of a rear stabilizing roller (cage or packer) whose position is modified by acting on cranks or on the third point of the lift. In addition to the depth adjustment, this roller ensures a crumbling and compacting effect of the soil.

-oscillating or reciprocating bar harrows: the adjustments are similar to those of the rotary harrow (equal lifting spindles, tilting to modify the inclination of the tines, frequency of the bars and working depth) **Fig 34.**

Combined tools, mainly used for surface cultivation, combine various tools on the same frame. Their main advantages are: working on large widths, better use of the tractor's power and, above all, a reduction in the number of passes.

The associated tools also make it possible to limit soil compaction by reducing the number of passes and to increase the speed of work. The most common combinations carry out preparation and seeding in one pass. With recent developments, it is even possible to combine ploughing, surface preparation and seeding. **Fig 35**

III. sowing and planting

The role of a seeder is to place the seed in the soil in order to plant a crop. It must allow :

- to provide a precise dose of seeds per hectare (or seeding rate) which is variable according to the size and weight of the seeds;

- to distribute the seeds evenly over the seed line;

- place the seeds at a regular and constant depth.

There are two categories of seeders:

- **The seed drills in line:** they allow to realize a continuous sowing (several seeds at the same time) on a line of sowing. They are mainly used for cereals and forage crops. They must therefore be versatile because the size of the seeds and the doses per hectare are very variable.

Table 1: Seeding rates for different row crops

Nature	Seeding rate (kg/ha)	Row spacing (cm)
Cereals	120 à 180	10 à 20
Soybeans	120 à 150	10 à 20

Rapeseed	2 à 5	10 à 30
Legume (forage)	2 à 10	10 à 20
Grass (forage)	10 à 20	10 à 20

Single-seed drills: they generally have one sowing unit per row which deposits the seeds one by one in the soil at regular intervals. They are reserved for weeded plants (maize, beet...). This type of seeder allows an appreciable saving of seeds.

The seeders are also classified according to their mode of proportioning or distribution which can be mechanical or pneumatic. Also the transport of the seeds to the ground can be mechanical (by gravity) or pneumatic.

1. Mechanical seed drills

They carry out a continuous sowing on a line and ensure the following functions **(see copy)**:

- grain storage in the hopper ;
- distribution by a mechanical dosing system;
- transport and implementation of the grain by a burying system;
- covering of the seed.

■ **Hopper**

The width of the hopper is equal to the working width. Its capacity is about 150 to 200 liters per meter of width. The agitator (horizontal axis) is animated by a rotary or alternating slow movement to ensure a regular descent of seeds without damaging them.

■ **Distribution system**

The mechanical distributions are made of a horizontal cylinder per row located under the hopper outlets. There are two models: the cylinder with grooves and the cylinder with lugs **(see copy)**. The adjustment of the grain flow is done by modifying the rotation speed of the grooves (flow proportional to the speed of advance DPA) and the useful length of the grooves.

■ **Transport and burying system**

Telescopic tubes (made of steel or plastic) ensure that the grain is lowered by gravity to the burying devices. This explains why the width of the hopper must be equal to the working width. **Fig 36**

- **Covering system**

Seed coverage is provided by a rear harrow with a single row of flexible tines. The depth of the tines must not be excessive in order not to modify the seeding depth.

- **Motion capture**

The drive is made through the wheels (carrier and drive) so that their output is strictly proportional to the area covered. Thus the quantity per hectare is constant whatever the working speed or via the tractor's power take-off. The main disadvantage of mechanical seed drills is the width of the hopper, which makes them cumbersome, especially from 4 meters. This explains the appearance of seeders with central hopper.

Table 2: Comparative study of the most used seed drills

Comparison criteria	Fluted seeders	Seed drills with lugs
Type of distribution	Forced	Accompanied/mixed/semi - forced
Variation of the flow	Truly proportional to the speed of travel	Variable according to the load of the seeds on the lugs
Regularity of distribution	Very good	Good
Crushing of seeds	**Risk to fragile grains**	no
Adaptation to the distribution of seeds of extreme size (very large or very small)	**Difficult**	Easy
Unfavourable influence of the slope on the distribution	Practically zero	**The spur delivers more downhill than uphill**
Easy to adjust	relatively easy	**More delicate (position of the pin, its speed of rotation, valve...)**
Purchase price	affordable	**Higher**

- **Optional equipment**

- the anti-crushing crutches of the shares ;

- trace erasers;

- Packing wheels and depth limiters;

- plotters ;

- the survey stakers.

- **Seeder adjustment**

- The farmer must fix a desired stand (number of plants/m^2) then from the weight of 1000 seeds (given for cereals), he must calculate the dose in kg/ha. The rate per ha of the seeder must be set at a fixed station:

- the adjustments are made according to the manufacturer's indications allowing approximately the choice of a correct adjustment for a given seed and flow rate;

- The frame is lifted with the help of the wedges and the seed is poured into the hopper;

- a tarpaulin (or better still a flow trough) is placed under the downpipes and the burying devices;

- the wheel circumference C (m) and l (m) the working width are measured. The area sown per wheel revolution is C x l ;

- we mark a point of the wheel and we make a test of flow, by making turn the wheel manually during 10 turns, by supervising the descent of seeds;

- the quantity of seeds that fell on the sack is weighed q (g) and the result is related to the ha to obtain the real flow rate Q (kg). If this flow differs from the desired flow, the test is repeated with a different setting.

(C x l x 10 trs x 10_{-4})ha .... (q x 10_{-3})kg

1   ha   .... **Q/ha = q / (C x l)**

❖ This adjustment can be made by simulating the sowing of one are. **1 are = 10^{-2} ha= 100 m^2**

The number of wheel revolutions required is thus **n = 100/(C x l)**

The seed obtained must represent one hundredth of the quantity to be brought to the ha.

❖ If the number of grains per m^2 is used, the operation should be as follows:

Seeding rate (kg/ha) = (No. of grains/m^2 x weight of 1000 grains (g)) / 100

❖ You can also check the flow rate by sowing a few metres on a flat surface and then count the number of grains that have fallen per linear metre and compare it to the number you wish to sow. The adjustment of the seeding rate is obtained from the following formula:

❖ **No. of grains/m^2 = seeding density (grains/m^2)/ No. of lines/m^2**

❖ **Seeding rate (grains/m^2) = desired stand (plants/m^2) / germination capacity**

The distance between the seed rows can be adjusted by moving the coulter mounting arms to the side.

The adjustment of the seeding depth can be done either by a mechanical or hydraulic centralized control, or by an individual control by spring in traction or in compression.

The track markers are usually adjusted in relation to the outer soil components. **(See copy)**

D = (L + E - V) / 2

L (cm) = working width (number of rows multiplied by the distance between the rows ;

E (cm) = distance between rows ;

V (cm) = front track of the tractor ;

D (cm) = distance between the outer coulter and the track disc.

The driving must be such that the speed does not exceed 8 km/h. and the headlands must be wide enough to allow a very good re-rowing of the seeder.

2. Pneumatic seed drills

This type of seeder has a central hopper. The space requirement is therefore reduced. It is thus possible to have seeders of great width (6 m and more) foldable on road. The seeding units are fixed on a bar whose ends are hydraulically foldable. They are also more easily used in combination with a soil cultivation tool (rotary harrow). Depending on the type of distribution, a distinction is made between :

-Pneumatic seed drills with centralized distribution: the seed drill has a centralized mechanical dosing system and then a pneumatic distribution and transport of the seeds to the seeding units. This principle was developed by the company Accord and is now used by many manufacturers (Kuhn, Vicon, Amazone...). A turbine driven by the PTO creates an air current in a vertical duct. The seeds are introduced there at the level of a venturi to facilitate their setting in movement. A conical top cover allows the seeds to be distributed over 360° and to feed the outlets evenly distributed around the perimeter. The air current then transports the seeds in rigid but flexible pipes until the organs of burying **(Fig 37)**.

-Pneumatic seed drills with multidistribution: these models also have a central hopper with a width smaller than the working width. It is located above the distribution system comprising as many cylinders with lugs or grooves as there are rows in the seeder. The flow rate is adjusted by changing their speed of rotation. Each distribution feeds a row. The grain is then transported by an air current due to a turbine to the burying devices **(Fig 38).**

Pneumatic seed drills are generally seed drills with mechanical metering and pneumatic transport. The principle of the metering system is the same as for mechanical seed drills. Their main interest is to be able to position the hopper independently of the burying devices. This is very useful for combinations of tools (soil cultivation tool + seed drill).

3. Single-seeders

For some crops, the distribution of seeds in the sowing rows according to equal and well determined spacings is sometimes necessary to obtain a suitable vegetative development and an optimal yield. These results are obtained with precision seeders or monograins, whose distributors are often designed to deliver the seeds one by one, each of them being separated from its neighbors by an identical distance. Precision seeding also saves seed and facilitates subsequent work.

The sowing units are mounted on the frame at a distance corresponding to the row distance. Each seeding unit consists of a hopper, a distribution system, and burying and covering elements. It thus ensures the sowing of a row.

> The distribution system allows the seeds to be picked up one by one before being dropped to the ground. The first models were equipped with a mechanical distribution system, consisting of a disc that picks up the seeds one by one. There are many models that differ from each other in the position and shape of the disc. **Fig 39, 40, 41, 42 and 43.**

NB: Mechanical distributions are simple, reliable and robust systems. They have two disadvantages:

- they impose the use of discs adapted to the seed used (size of the seed);

- the speed of advance of the seeder must be low (max 3 to 4 km/h) to avoid the misses and the doubles (sowing of vegetables and sugar beets). In field crops, pneumatic distributions allow a higher working speed (8 to 10 km/h).

- The pneumatic seed drill must be equipped with a turbine driven by the PTO. It supplies an air flow downstream and therefore an air pressure and upstream a suction or a vacuum. Depending on the model, the suction or vacuum can be used separately or together for the operation of the distributor **(Fig. 41).**

  ■ **Burying devices**

Each sowing unit is equipped with :

- a deflector or front roller (levelling roller);

- an opening device to make a furrow at the desired depth where the seed falls. This is usually a trailing coulter (sowing coulter);

- one or two wheels carrying the sowing unit which also serve to adjust the sowing depth;

- a sealing and covering device consisting of a sealing wheel, support wheel, covering skimmer or an oblique deflector (tooth).

The seeder can be equipped with a microgranulator whose role is to spread an insecticide or fungicide product (in powder or microgranules) on the seeding line at the same level as the seed and fertilizers whose purpose is to deposit in the soil by burying teeth, a fertilizer known as "starter" which will promote the start of the plants, on the side of the row and deeper in order to force the roots to go down in the soil.

Packs and levels the soil in front of the coulter to ensure a more regular penetration	Front roller or levelling roller
Eliminates clods and surface stones to facilitate ploughing	Clump breaker or lump breaker device
Ensures the partial covering of the seeds and especially the compaction of the soil	Pressure wheel
Finish filling the furrow with fine soil that limits evaporation	Covering shaver
Cup in the furrow to press the seeds into the soil	Support roller (beet)
Opens the furrow in which the seeds are deposited	Sowing Company

■ **Practice adjusting the flow rate and controlling the seeding rate (stationary).**

For a chosen disk and speed, the manufacturer indicates a density that must be checked.
-Place a bucket under each coulter and rotate the drive wheel as much as necessary to cover a distance of 100 m (no. of revolutions = 100/C) with C = circumference of the wheel (m).

Q = the dose to be sown per ha (kg) ;

L = the length of the lines per ha (m);

E = line spacing (m) ;

100 ....mq

L (m)................Q

Q = (L x q) / 100 or L = 10000/E so **Q = (100 x q)/E**

To obtain a regular seeding, the choice of the working speed is essential and very variable according to the distribution systems or the brands used: the recommended speeds are from 5 to 7 km/h for the pneumatic seeders, from 4 to 5 km/h for the mechanical seeders; they can even go down to 3 km/h when the ground is stony or too cloddy.

Excessive working speed leads to :

- a decrease in the quantity sown and in the effective density ;

- some irregularity on the line;

- burying the seeds at uneven depths.

In order for a single-seed seeding to be done correctly, three factors must be met:

- respect of the recommended density;

- a regular distance between the seeds on the sowing line;

- a constant depth of seed burial.

4. Potato planters

These materials must allow the planting of different sizes of seed potatoes, at regular distances and depths, without damaging the sprouts. The rules to respect for the setting of the tubers are :

- Respect on the line of the regular intervals;
- Constant planting depth ;
- Same spacing between rows;
- Sufficiently loose soil and not too wet.

a. Hand-fed planters

The systems still in use are, on the one hand, vertical rotary distributors (notched cylinder or bucket chain) and, on the other hand, horizontal rotary distributors (compartmentalized tray). **Fig 44, Fig 45, Fig 46**

b. Planters with automatic feeding

The distribution system is fed without manual intervention. Several dispensing systems are classified into three categories:

- Chain or belt systems with buckets **Fig 47,**

- Aligner conveyor systems **Fig 48, 49**

- The disc and finger puller systems **Fig 50**

1.3. Constituent elements

- **The frame**

Profiled, of the "tool holder" type, it allows the displacement of the planting elements in order to adjust the spacing between the rows (between 55 and 90 cm).

- ■ **The hopper**

This is the bin, whose capacity varies according to the number of items it can serve.

- ■ **The distribution system**

There are manual and automatic feeders. The manual distributor can be vertical rotary (the tubers are deposited one by one in notches or cups, held in place during rotation and released at ground level) or horizontal (the tubers are deposited on a compartmentalized tray rotating in a drum pierced by a skylight that opens onto the drop tube). The automatic distributor can be made up of a belt or bucket chain (the tubers are taken from the lower part of the hopper), or a system with a conveyor-aligner (the tubers are aligned on a belt with an adjustable rotation speed. This system is used in particular with pre-sprouted plants), or a system with discs and extracting fingers.

- ■ **The grounding and covering organs**

A height-adjustable ridging share opens the furrow in which the plants will be placed. The regularity of the working depth is ensured by a mounting of the shares on a parallelogram system and a support wheel. The covering is made by two discs.

The valves are always driven by a chain or gearbox from the drive wheels.

1.4. Settings

- **adjustment of the plant spacing in the line:** the distributor is equipped with a series of gears that allow to modify its rotation speed, and therefore the rhythm of the deposits for the same advancement speed.
- **adjustment of the distance between the rows:** on machines with several rows, the elements can be moved laterally on the frame, to vary the distance between the rows.
- **adjustment of the burying and covering**: the furrow opening shares are fixed to the frame by brackets which allow to quickly change their depth. The height, distance, inclination and ground pressure of the covering discs can be adjusted.

Chapter 3: Study of the tools for the maintenance of the crops Fertilization and treatment materials

The application of plant protection products and liquid fertilizer is done with sprayers, while solid forms are applied with fertilizer spreaders. Finally, slurry and manure spreaders have been designed for the spreading of various farm fertilizers.

1. Fertilizer distributor

Fertilizer spreaders are used to spread a specific quantity of solid fertilizer on a part (localized spreading) or more often on the whole (generalized spreading) of the plot.

1.1. Constituent elements

The basic elements of any fertilizer distributor are:

- a system of connection with the tractor, the distributors being able to be carried, semi-mounted or trailed,

- a hopper in which the fertilizer particles are stored,

- a distribution system to distribute the fertilizer on the ground. In order to remain constant, the spreading rate must be proportional to the advance, thanks to a drive from one of the wheels of the fertilizer distributor.

- a transfer system and output devices are present on some types of distributors

1.2. The different types of fertilizer spreaders and how they work

1.2.1. Centrifugal fertilizer spreaders

These are the most widely used devices (for generalized spreading) because of their low cost, simplicity and good performance. On the other hand, they are rather delicate to adjust precisely. They are usually mounted (hopper capacity: 400 to 2000 L) or semi-mounted (hopper capacity: up to 10 000 L). **Fig 1**

As they leave the hopper, the fertilizer particles hit 1 or 2 rapidly rotating paddle discs which throw them out by centrifugal force. The distribution organs are fed either by gravity or by a hydraulically driven screw or by a belt driven by a wheel running on the ground. The movement of the discs (in the opposite direction) is generally ensured by the power take-off. **Fig 2, 3** The farmer can act on :

- **the flow rate**: by modifying the opening of the dosing hatch, or by varying the rotation speed of the screws or the feeding belts.

- **the working width**: by changing the height and inclination of the discs, the orientation of the disc blades, the point of arrival of the fertilizer on the discs.

In order to accurately calibrate the device at each fertilizer change, control kits are available to check the actual flow rate and spreading width.

1.2.2. Pneumatic fertilizer spreaders

They can be mounted, semi-mounted (hopper capacity: 600 to 800L) or trailed (capacity up to 6000L). These devices are more expensive and more delicate to maintain, but offer a great precision of dosage even on large widths and whatever the qualities of the fertilizer. **Fig 4**

The fertilizer granules are carried by an air stream to the diffusers where they are distributed on the ground. The distribution system consists of spiked rollers that introduce the exact dose of fertilizer into each transport tube. Through these tubes, which are 0.25 to 1.50 m apart and arranged parallel on the boom, the product is transported by an air flow produced by a turbine and is conveyed to the outlet openings called diffusers.

The farmer may pay:
- fertilizer dosing (action on the rotation speed of the dosing rollers),
- the fan flow (action on a shutter),
- the orientation of the deflectors of the diffusers (upwards or downwards) **Fig 5, 6**

1.2.3. Fertilizer distributors

By adapting location sleeves on the diffusers, the classic pneumatic distributors can be used for localized fertilization. They deliver fertilizer in continuous bands a few centimeters from the plants. Locating distributors can also be used in combination with a conventional seed drill or a precision seed drill. In this case, one or two hoppers are connected to the seed drill and feed each seed line through the drop tubes.

2. Manure spreader

Manure spreaders, semi-mounted trailers equipped with a spreading device, are, as their name indicates, designed to spread manure on a plot of land. These manure applications are reasoned according to the needs of the crop and the composition of the manure (OM and N contents).

2.1. Components and operation

The spreaders consist of a storage tank, a feeding system and a spreading system.

■ **Tanker**

The tanks are semi-mounted, fixed on a one or two axle chassis, and can hold between 3 and 15 T of manure.

■ **Power supply system**

Metal bars fixed on endless chains make up a moving floor allowing the manure to be advanced towards the distribution organs.

■ **Spreading system**

Rear discharge spreaders (Fig. 7, 8): horizontal or vertical reels equipped with dispersing elements (tines, claws, crenellated discs, helical screw...) are rotated by the tractor's power take-off and have the role of shredding and spreading the manure.

The system can be completed with a spreader system or a spreading table to increase the spreading regularity and the spreading width. Vertical reel spreaders provide a larger spreading width (up to 8 m). **Fig 9 and 10**

Lateral spreaders: They are designed for spreading semi-liquid manure (thick slurry, poultry droppings). There are systems with flail chains and rotor systems.

- **Flail chain system:** A rotor driven by chains moves progressively from the top to the bottom of the box while throwing the material. There are horizontal systems in which a moving bottom brings the material to the front as it is projected.

- **Rotor systems:** a moving floor brings the manure to the front of the trailer at a tine rotor that throws the material sideways through a swivel outlet nozzle.

3. Slurry spreader

Slurry spreaders, also called "slurry tankers", are used to mix the slurry in the pit, to draw it out and to spread it. Some devices are even equipped with a burying system that can be fitted to the back of the spreader. **Fig 11 3.1 Components**

■ **Tanker**

It is generally cylindrical in shape, can store between 3,000 and 20,000 L of slurry and is resistant to corrosion and pressure variations during loading and spreading. The rear wall can be opened or removed for cleaning the interior.

■ **Pump**

It is a lobe or vane pump that can work alternatively as a compressor (when spreading, the manure is pressurized) or as a vacuum pump (to create the air vacuum before filling the tank) thanks to a reversing valve system. The pump only works on air and never comes into contact with the slurry.

■ **Spreading system and incorporation system**

Sheet diffusers :

When the slurry leaves the tank, it is sprayed onto the ground in a layer by means of a deflector (directed upwards or downwards), a vertical serrated disc, a horizontal paddle disc or a pivoting distributor.

Ramps with trailing pipes:

This device allows a spreading very close to the ground and thus limits the risks of losses by volatilization. These foldable booms, which can handle widths of 12 to 21 m, are equipped with numerous parallel flexible pipes spaced 25 cm apart, through which the slurry flows to the ground. **Fig 12**

- **Landfill system**

The ploughs for cultivated land: resemble the tine cultivators of the "chisel" type. The advantage of this device is the possibility to dethatch the crops while adding slurry without emitting unpleasant odors. **Fig 13**

Grassland ploughs: are equipped with furrow-opening discs or knives and place the manure at a depth of 10 to 15 cm in the soil. **Fig 14**

Table 3: Comparison between centrifugal and sheet feeders

Elements of comparison	D. centrifugal	D. on the tablecloth
Fertilizer dispersal method	Effect of centrifugal force	With free or sometimes forced exit
Spreading width	Large (max 24 m)	0.9 to 4 m
space requirement	low	Important
weight	slight	heavy
Cleaning and maintenance	easy	difficult
Purchase price	The cheapest	More expensive
Work capacity	large	Smaller
Settings	delicate	Easy
Effect of wind, slope	important	Less important

- **Advantages and disadvantages of pneumatic systems**

+ Unaffected by wind, shaking and slope of the land

+ Homogeneous distribution

+ Versatile systems

+ Certain ease of emptying and dismantling before washing with a water jet.

Height up to about 3 m, which makes it difficult to fill the hopper

_ cumbersome and heavy spreading ramp to fold up

High purchase price.

4. Sprayers

A sprayer is used to spread a certain dose of plant protection products or liquid fertilizer in the form of fine droplets by distributing them evenly over a target (the soil, the crop stand, weeds, etc.).

Sprayers can be trailed, semi-mounted or mounted (in front or behind the tractor) and some are self-propelled. **Fig 15**

4.1. Constitution and operating principle

▪ Tank

Product storage of variable capacity (from 200 to 6000L), with self-propelled sprayers having the highest capacity, ahead of trailed and mounted sprayers respectively.

▪ Distribution system

It is a pump whose role is to :

- to take a given quantity of liquid from the tank and to deliver it to the liquid transfer system under a given pressure,
- stir the liquid in the tank,
- Fill the sprayer (on small units, as a second pump is used for larger units).

▪ Liquid transfer system

The boom corresponds to the pipe supporting the nozzles and is fixed on a frame or boom support (itself attached to a frame) which allows the articulation of the boom and its height adjustment.

The ramps are of variable width (from 6 to 40 m) according to the capacity of the tank, the self-propelled ones having thus ramps of important widths.

▪ System for dividing the liquid into droplets

Nozzles, calibrated orifices distributed along the boom, control the shape of the jet, its distribution on the target, the size and number of drops. **Fig 16**

Droplet division can be generated by:

- **Liquid pressure**: the liquid arrives under pressure in the nozzles and expands at atmospheric pressure, causing it to split into drops. This process is mainly used to treat low crops.
- **Air pressure**: An air stream at the nozzles divides the liquid arriving at low pressure into drops.
- **centrifugal force**: The liquid arrives tangentially on a disc rotating at high speed and is broken up into drops by centrifugal effect

▪ Drop transfer system

The drops must acquire enough kinetic energy to reach their target.

5. transport by projected jet: the drops are transported thanks to the kinetic energy stored during the division of the liquid into drops.

6. transport by jet: an air flow carries the drops to their target.

❖ Control system

DPA (proportional flow control) or **DPM** (proportional flow rate to tractor engine speed, i.e. pump speed) are frequently installed to control and maintain a constant spray rate. The **DC** constant flow control device provides a constant boom flow regardless of the pump flow (with a pressure regulator).

There is also the proportional flow control **(CPA) system**, where a pump in a constant flow water circuit provides proportional dosing of the spray solution.

- ❖ **The different types of sprayers**
- ❖ **Liquid pressure and jet sprayers**

The potential energy of the liquid, pressurized by the pump, is used to transform it into drops (division by liquid pressure) and to transport them to the target (transport by jet) through nozzles. **Fig 17,18,19, 20**

- ❖ **Centrifugal sprayers**

They are used in field crops to work at low volumes (50 to 100L/Ha) and use centrifugal force and spray transport. **Fig 21**

- ❖ **Pneumatic sprayers**

An air current divides the jet and carries the formed drops. These devices are particularly used in viticulture. **Fig 22**

Chapter 4: Study of mowing and harvesting equipment

The harvesting of grains (cereals, oilseeds) is done with a combine harvester. The harvesting of roots and tubers is done with different machines (digger, leaf stripper, stripper...) which can be used one after the other or simultaneously in combined form. The machines used to harvest fodder are also very varied and depend on the type of harvest (dry or wet).

1. Mower

Mowers perform the first step in harvesting forages: cutting. They cut the forage at the base of the stalks as cleanly as possible and leave the product loose on the ground or, more often, in rows called windrows. The mowing units can be placed in a front, rear or even side position and are driven by the tractor's power take-off.

1.1. Constituent elements

1.1.1. Cutting system

Cutting by sectioning (or "cutting with support"):

It is made by a cutter bar which consists of :

- a cutting blade or saw: a moving part with cutting sections that moves in an alternating rectilinear motion.
- a blade holder or "support bar": a fixed part equipped with fingers in which the cutting blade slides. It can be 3 to 4 m wide and slides on the ground thanks to 2 shoes. One end of the blade holder supports the swath board, which separates the standing crop from the cut crop.

  Fig 1

Cutting by laceration ("cutting without support"):

- This cut, less frank than the cut by sectioning, is carried out by rotary elements (drums, or plates) provided with knives. **Fig 2, 3**

1.1.2. Packaging system

Conditioning the stalks means crushing them in order to accelerate the evaporation of the water they contain (up to 80%), which is unfavorable to their good conservation. Mowers are usually equipped with a conditioning device that can consist of either 2 ribbed rollers that bend and break the stalks **Fig 4**, or a finger rotor that catches the cut fodder and pulls it against a comb **Fig 5**. During this operation the leaves must not be altered.

1.2. Different types of mowers

- **Reciprocating mowers (with cutter bar)**

  They perform a cutting by sectioning. Although they allow a clean cut, their working speed is relatively slow (5 to 8 Km/h) and they are not adapted to stony or badly levelled grounds...

- **Rotary drum mowers**

  They have drums as cutting elements and make a cut by laceration. These machines have supplanted the previous ones because of their high working speed (12 to 15 Km/h) without jamming.

- **Mower conditioners**

  The mowers can be equipped with a conditioning system to perform the cutting and conditioning in one operation.

- **Mower conditioners windrowers**

  These machines perform mowing, conditioning and swathing at the same time. The self-propelled front-cutting versions, equipped with a comb reel, are used on large sites because of their power, which allows a very wide cut at high speed.

1.3.   Settings

- cutting height: it is obtained by action on the shoes for the alternative mowers, or by action on a screw crank (for the rotary mowers).
- orientation and length of the swath (for mower conditioners).

2. Forage Harvester

Silage harvesters or forage harvesters are used to harvest green forage (grasses, legumes, corn) or hay (grasses, legumes that have been previously mowed and conditioned). The product chopped by the forage harvester is then compacted and stored in an airtight place to be transformed into silage by fermentation.

2.1.   Different types and principles of operation

- **Flail foragers**

Generally used for green harvesting, these are the simplest machines in design. Flails rotating around a rotor cut the forage by laceration and send it to the loading trailer via an ejection nozzle. A distinction is made between single-cutting machines (spoon-shaped flails) and double-cutting machines (L-shaped flails, supplemented by a rotary chopping system), which provide a finer chop.

- **Fine cut forage harvesters (knife harvesters)**

These machines, generally self-propelled, offer a high output (more than 100 T of processed fodder per hour), a great precision and a high fineness of chopping (4 to 40 mm), conducive to the success of the silage. They are composed of : **Fig 6**

S **A harvesting head**

It consists of either :

■ a cutter bar (identical to that of the mowers) for direct cutting, - a pick-up (transverse bars with teeth whose ends are linked to rotating discs) to pick up previously cut and swathed forage, - corn harvesting spouts (harvesting corn-forage "plant-identity").

S **A feeding device**

The crop flows from the harvesting head to the chopper by means of two stages (one fixed, the other mobile) of grooved rollers or bar chains whose speed of rotation influences the length of the strands. The result is a regular, compressed strip that is easy to chop.

S **A hash system**

At the exit of the feeding system, the forage comes into contact with chopping drums (transverse rotating cylinders equipped with cutting blades, brushing against a tangential counter-knife)

S **A wind tunnel**

On high capacity machines, a fan causes the chopped product to be ejected into the storage hopper or the following trailer.

3. Combine

The combine harvester is used to harvest cereals and oilseeds. It automatically and simultaneously cuts the inflorescences, threshes the grains and cleans them to extract them from the husks. Combine harvesters are self-propelled and the engine power usually varies from 200 to more than 300 HP. The main characteristic of a combine is its hourly output (from 80 to 250 Q/Ha).

3.1. Constituent elements

3.1.1. Basic elements

A combine harvester includes:

- a cutting system (cutter bar, lifters, dividers, reel, feed screw)

- a threshing system (threshing drum, concave, stone trough, straw puller)

- a separation and cleaning system (shakers, grain table, cleaning box

- a storage system (hopper)

3.1.2. Additional elements

- **straw chopper**: located at the exit of the straw walkers, this horizontal axis rotor equipped with knives and counter-knives, chops the straw before spreading it.

- **straw spreader**: located behind the shakers, this vertical axis rotor disperses the remains of stalks and inflorescences, to facilitate their later burial.

- **corn pickers**: they replace the cutting table, and allow to pick only the cobs, leaving the stalks on the spot. The thresher has therefore a much smaller volume of material to process. These corn grain harvesters are called "corn-sheller". Cornpickers can also be used for harvesting unshucked ears (but this method tends to disappear).

- **harvesting equipment for oilseeds**: modifications to the reel and presence of adapted trays for sunflower (limiting the loss of heads); cutting dividers and extended cutting table for rapeseed (limiting the loss of seeds).

3.2. Technological innovations

To ensure the high flow rates of modern machines, various devices have been developed:

- the rotary separators (drums placed after the straw puller) are almost equivalent to a chain of several beater/counter beater systems;
- rotary shakers: these rotors placed above the grids improve the separation;
- Axial flow threshing: the threshing drum is arranged longitudinally, so that the harvested material is forced to run along the length of the threshing drum at a helical pitch, which increases the efficiency of the threshing process. This system is only used on farms where the straw is not pressed, as it crushes the straw considerably.

3.3. Settings

Depending on the type of crop, we can act on different parameters:

- Cutting height;
- Beater/counter-beater spacing and beater rotation speed;
- Interchangeable or adjustable opening grids that can be adapted to the size of the seeds;
- Ventilation power.

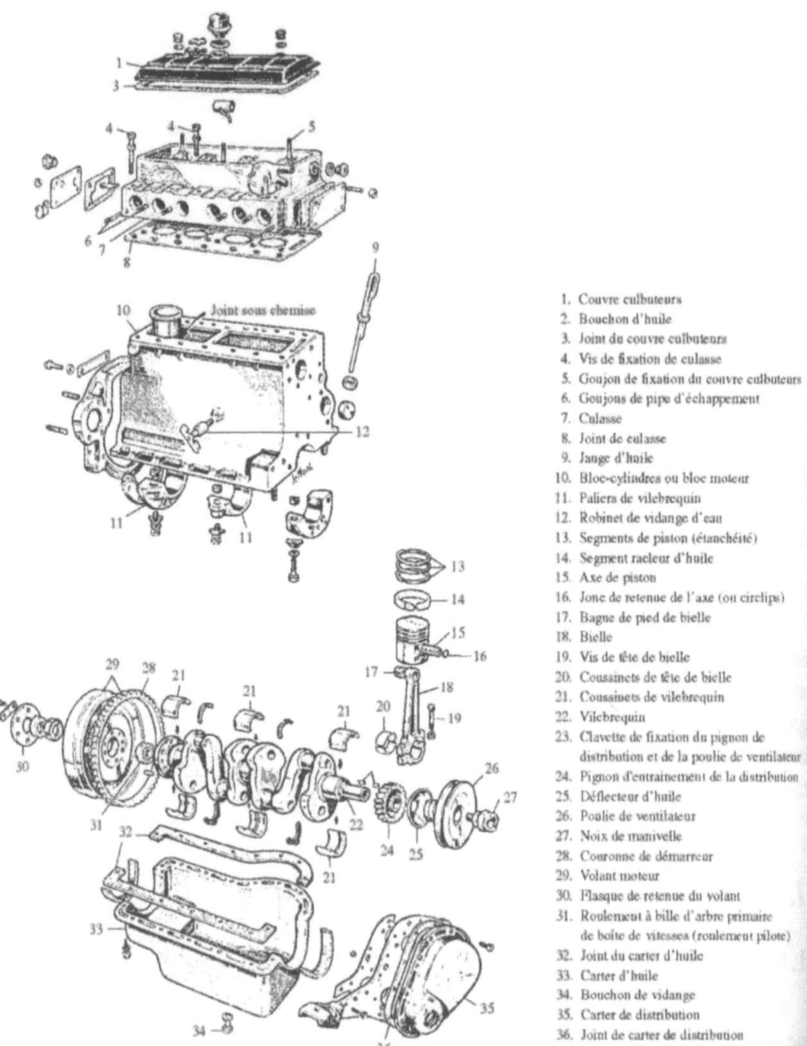

1. Couvre culbuteurs
2. Bouchon d'huile
3. Joint du couvre culbuteurs
4. Vis de fixation de culasse
5. Goujon de fixation du couvre culbuteurs
6. Goujons de pipe d'échappement
7. Culasse
8. Joint de culasse
9. Jauge d'huile
10. Bloc-cylindres ou bloc moteur
11. Paliers de vilebrequin
12. Robinet de vidange d'eau
13. Segments de piston (étanchéité)
14. Segment racleur d'huile
15. Axe de piston
16. Jonc de retenue de l'axe (ou circlips)
17. Bague de pied de bielle
18. Bielle
19. Vis de tête de bielle
20. Coussinets de tête de bielle
21. Coussinets de vilebrequin
22. Vilebrequin
23. Clavette de fixation du pignon de distribution et de la poulie de ventilateur
24. Pignon d'entrainement de la distribution
25. Déflecteur d'huile
26. Poulie de ventilateur
27. Noix de manivelle
28. Couronne de démarreur
29. Volant moteur
30. Flasque de retenue du volant
31. Roulement à bille d'arbre primaire de boîte de vitesses (roulement pilote)
32. Joint du carter d'huile
33. Carter d'huile
34. Bouchon de vidange
35. Carter de distribution
36. Joint de carter de distribution

Fig 1. exploded view of the different components of a four-stroke engine

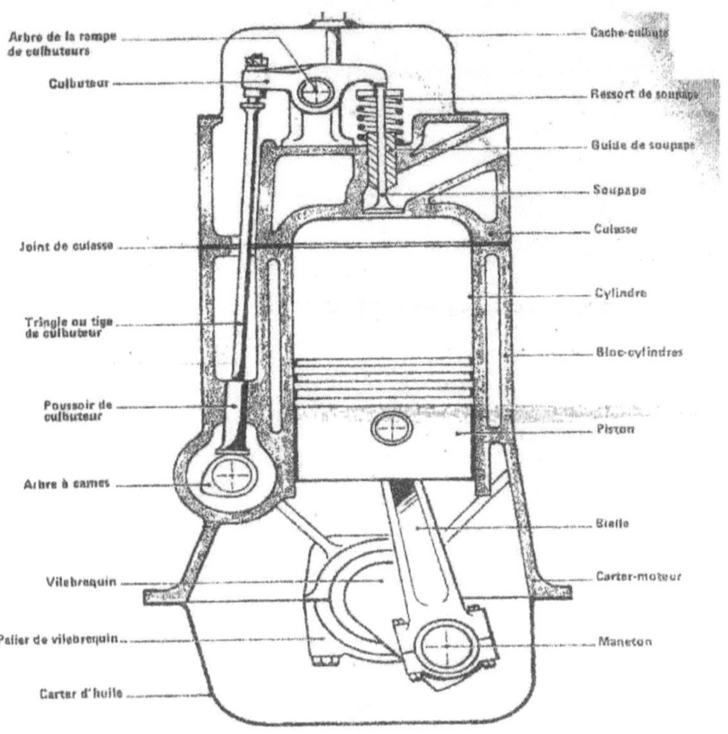

Fig 2. Components of a four-stroke engine

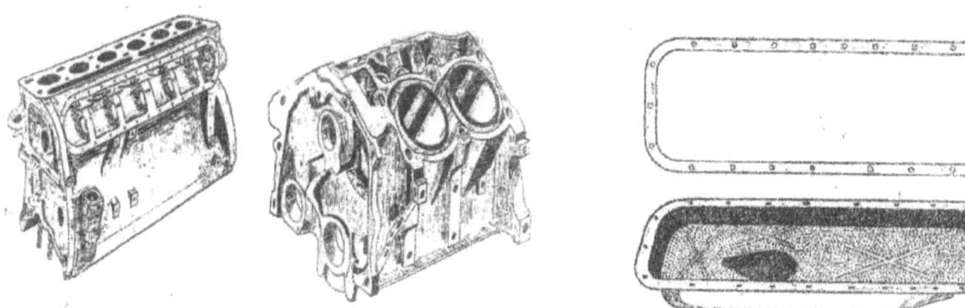

Fig 3. Cylinder block   **Fig 4. Lower housing or oil tank**

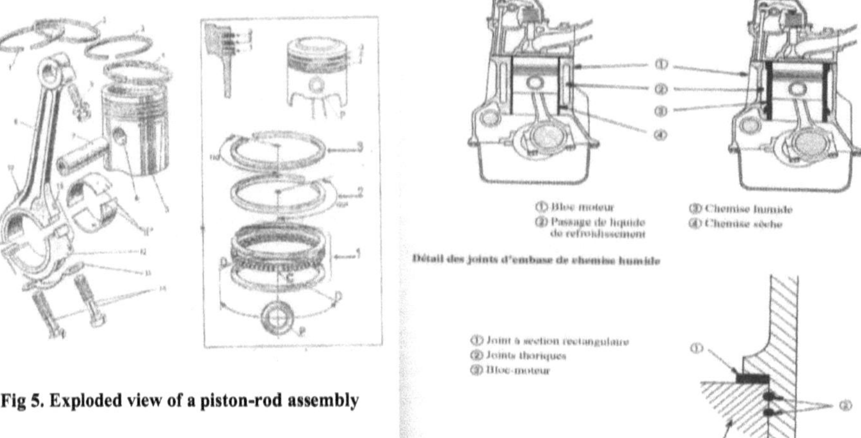

Fig 5. Exploded view of a piston-rod assembly

Fig 6. Dry and wet jacket of a water-cooled engine

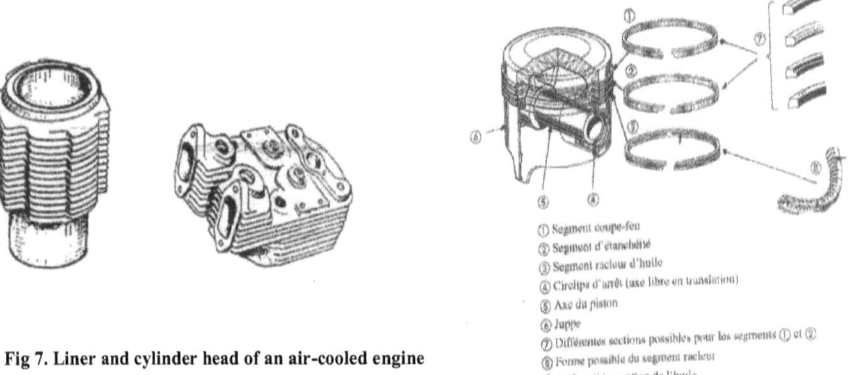

Fig 7. Liner and cylinder head of an air-cooled engine

Fig 8. Different rings mounted on a piston

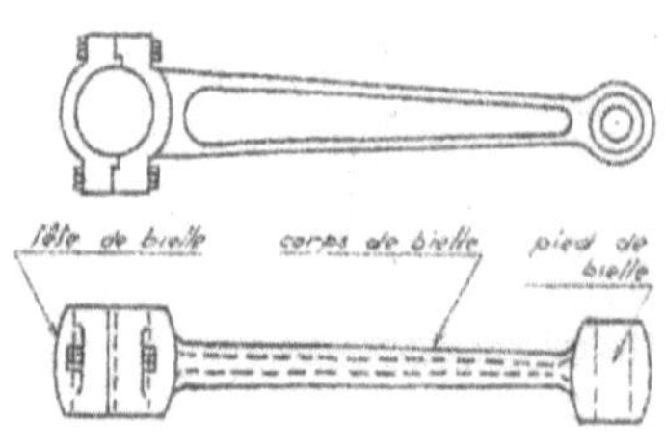

Fig 9. Elements of a connecting rod

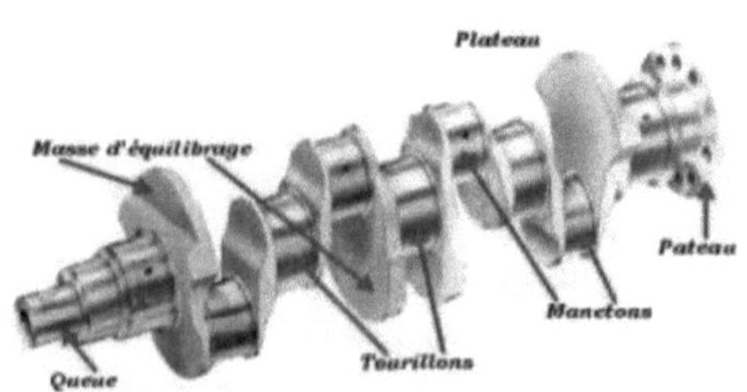

Fig 10. Crankshaft

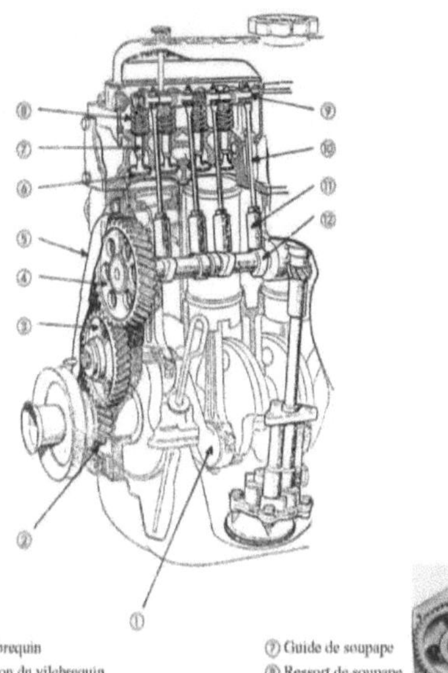

① Vilebrequin
② Pignon du vilebrequin
③ Pignon intermédiaire de distribution
④ Pignon de l'arbre à cames
⑤ Carter de distribution
⑥ Soupape

⑦ Guide de soupape
⑧ Ressort de soupape
⑨ Culbuteur
⑩ Tige de culbuteur
⑪ Poussoir
⑫ Came

Fig 11. View of a lateral camshaft

$$V = \pi \times \left(\frac{alésage}{2}\right)^2 \times course$$

Fig 12. Motor characteristics

① Pignon du vilebrequin
② Pignon intermédiaire supérieur
③ Pignon de la pompe d'injection
④ Repères de calage
⑤ Pignon de l'arbre à cames
⑥ Pignon intermédiaire inféri
⑦ Pignon de la pompe à huile
⑧ Emplacement des arbres d'équilibrage (non présent sur la photo)

Fig 14. Sprocket distribution

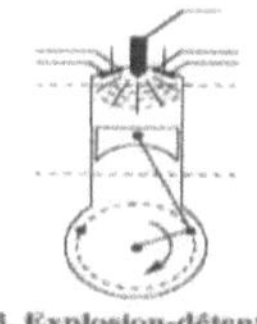

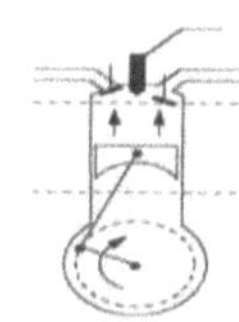

① Piston
② Bielle
③ Vilebrequin
④ Pipe d'admission
⑤ Soupape d'admission
⑥ Injecteur
⑦ Soupape d'échappement
⑧ Pipe d'échappement

Fig 13. Cycle of a four-stroke engine

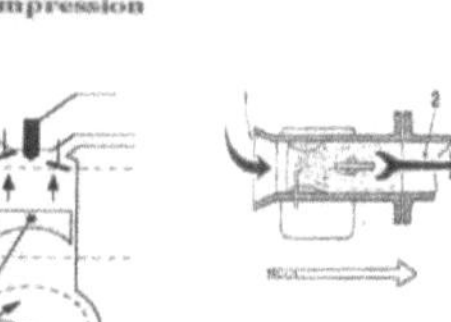
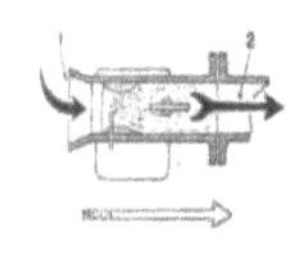
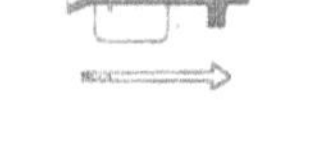

Fig 15. Principle of carburation

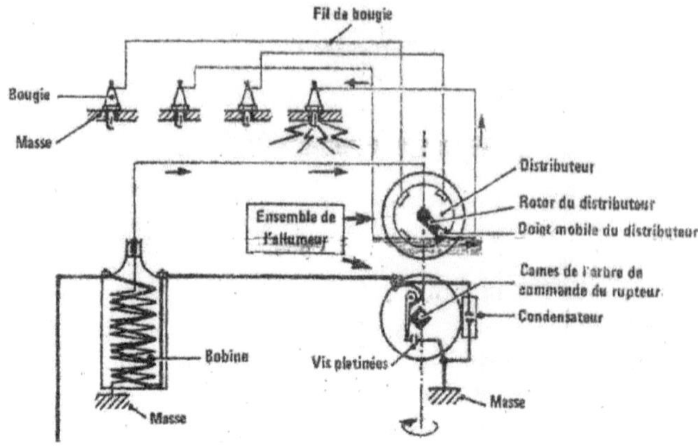

Fig 16. Ignition circuit of a four-stroke gasoline engine

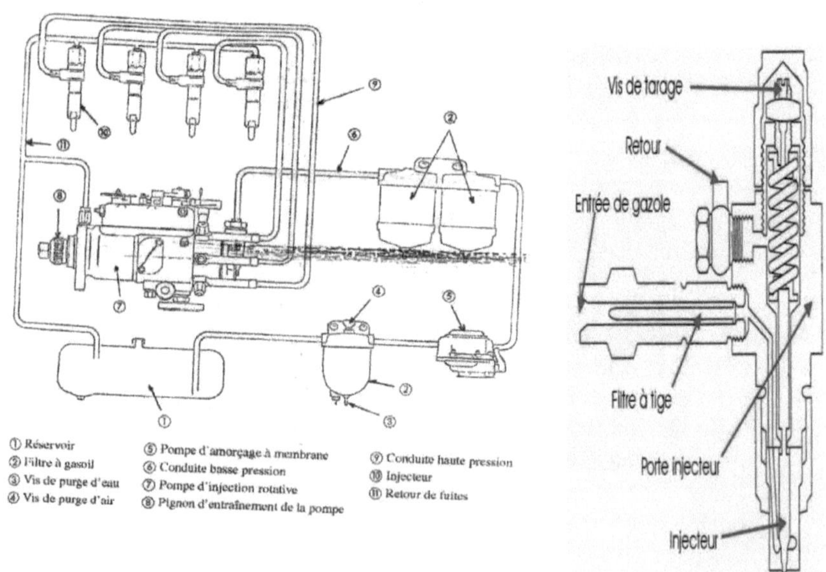

Fig 17. Fuel supply

Fig 18. Components of an injector

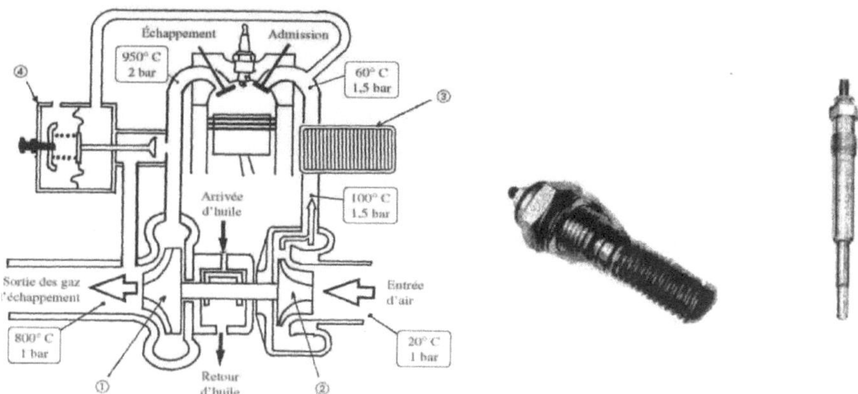

Fig 20. Diesel engine heater plugs

Fig 19. Principle of the turbo compressor on diesel engines

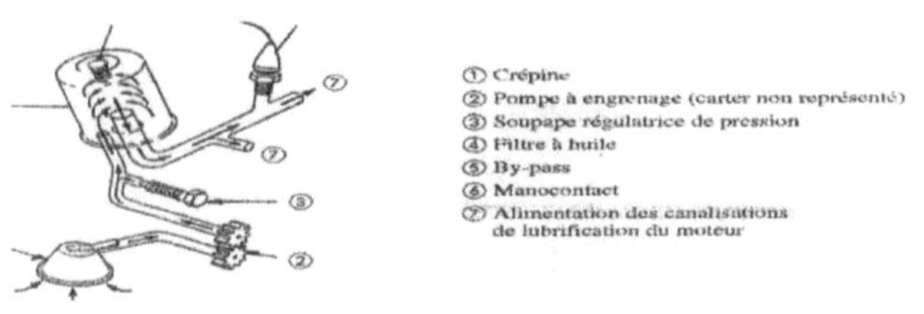

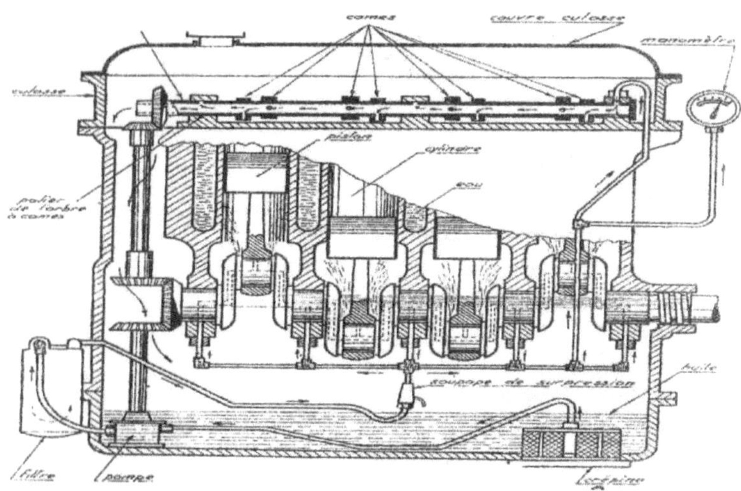

Fig 21. Lubrication circuit of a four-stroke engine

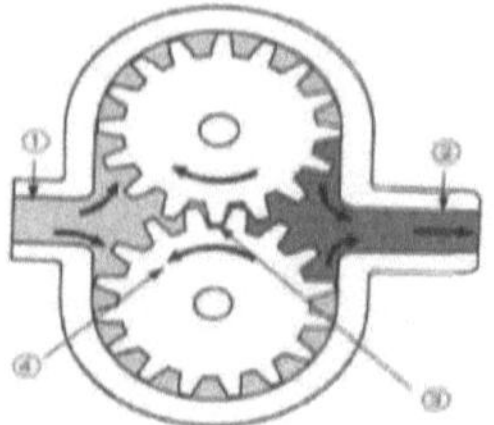

Fig 22. Operating principle of a gear oil pump

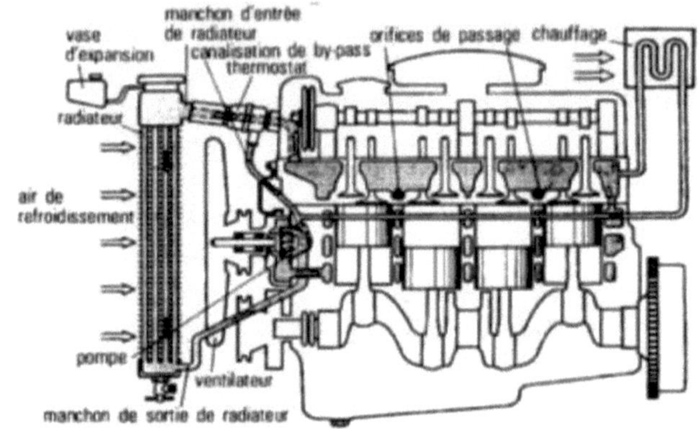

Fig23.Watercooling circuit

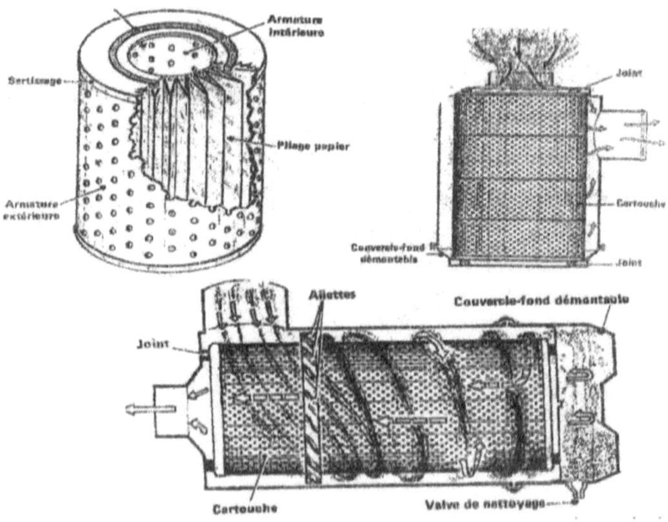

Fig 24. Dry Cartridge Air Filter

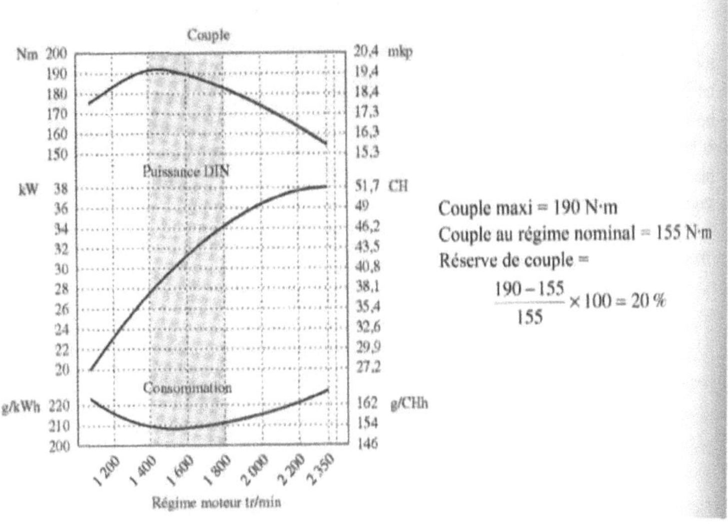

Couple maxi = 190 N·m

Couple au régime nominal = 155 N·m

Réserve de couple =

$$\frac{190-155}{155} \times 100 = 20\ \%$$

Fig 25. Performance curves for a 52 hp DIN engine

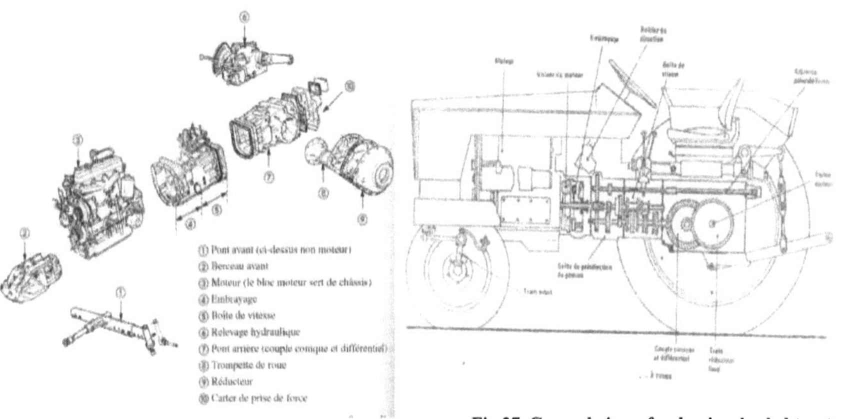

Fig 27. General view of a classic wheeled tractor

Fig 26. Main components of a wheeled agricultural tractor

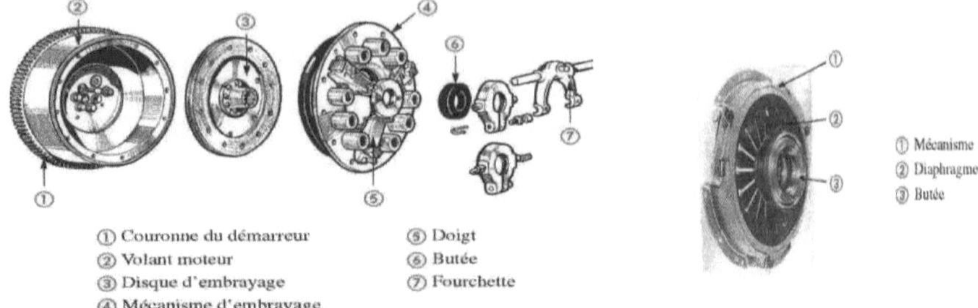

Fig 28. Main parts of a disc clutch   **Fig 29. Diaphragm clutch mechanism**

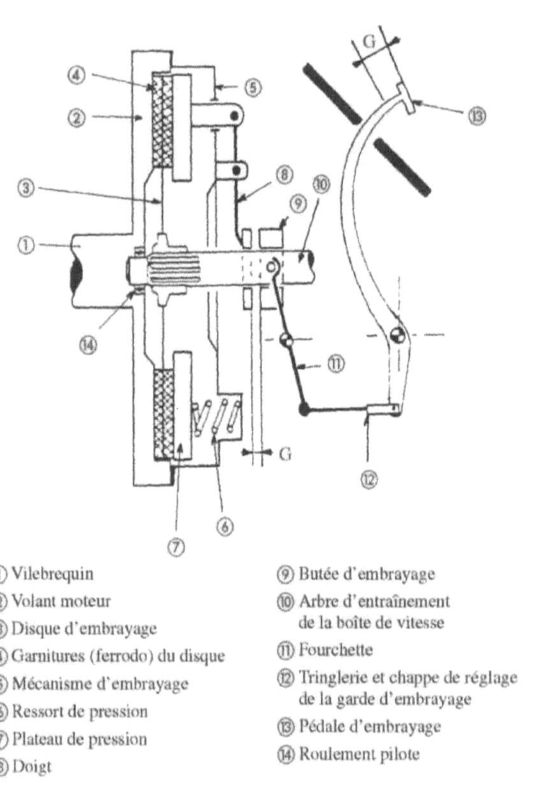

① Vilebrequin
② Volant moteur
③ Disque d'embrayage
④ Garnitures (ferrodo) du disque
⑤ Mécanisme d'embrayage
⑥ Ressort de pression
⑦ Plateau de pression
⑧ Doigt
⑨ Butée d'embrayage
⑩ Arbre d'entraînement de la boîte de vitesse
⑪ Fourchette
⑫ Tringlerie et chappe de réglage de la garde d'embrayage
⑬ Pédale d'embrayage
⑭ Roulement pilote

G : Garde d'embrayage

Fig 30. Principle of a single-plate clutch

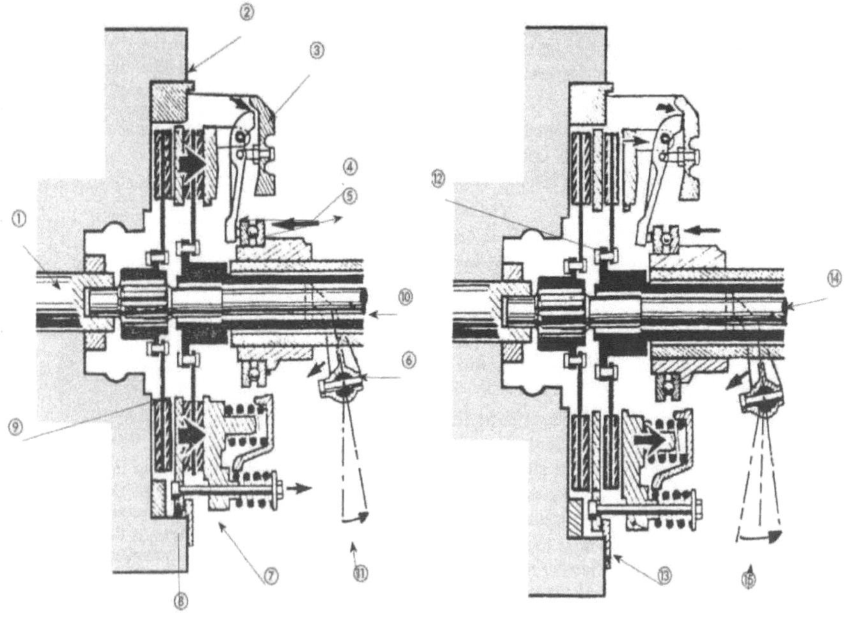

Fig 31. Double acting clutch

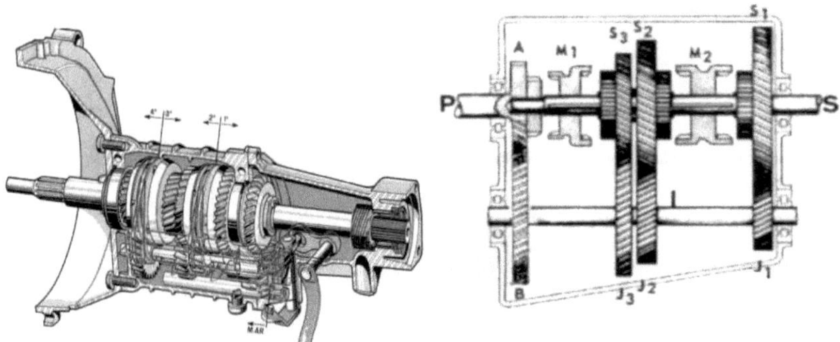

Fig 32. Elements of a gearbox

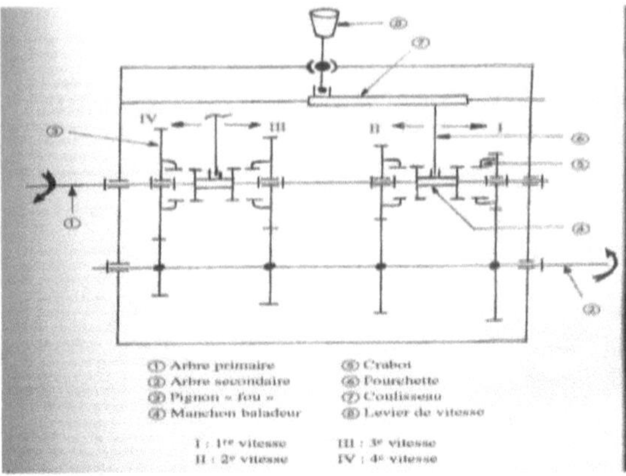

Fig 33. Block diagram of a four-speed gearbox

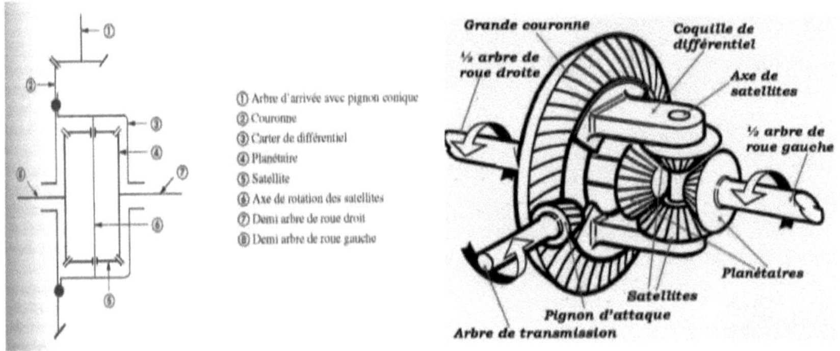

Fig 34. Schematic representation of a differential

Fig 35. Operating principle and main parts of a differential

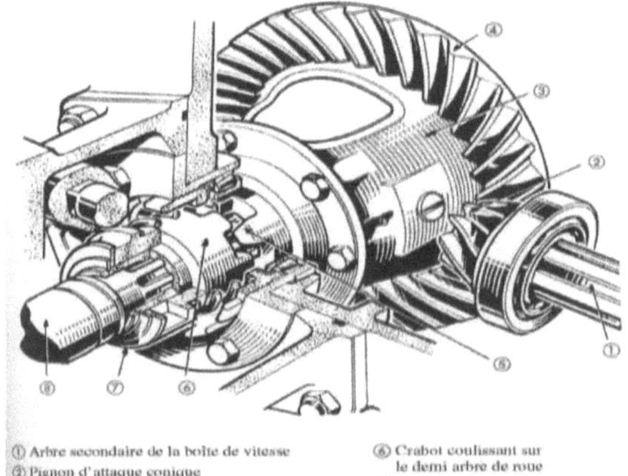

Fig 36. Mechanical differential lock

Study of the tools for setting up the crops

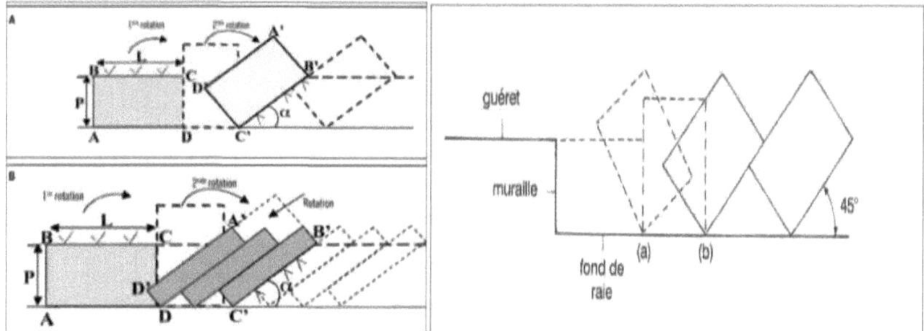

Fig 1: Schematic diagram of the principle of the reversal model of the strip of land cut by the plough. The cut strip is represented in section in a plane perpendicular to the direction of advance of the plough. The movement includes two successive rotations (a), then a translation (b) which simulates the collapse of the soil at the bottom of the furrow, after turning over (example with two shearing planes cutting the strip). A', B', C', D': position of points A, B, C and D after turning; P = ploughing depth, L = ploughing width.

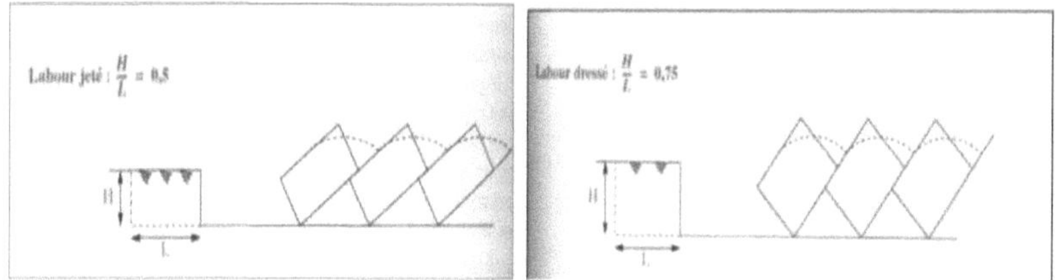

Fig 2: The two types of ploughing at the level of the ploughed strip

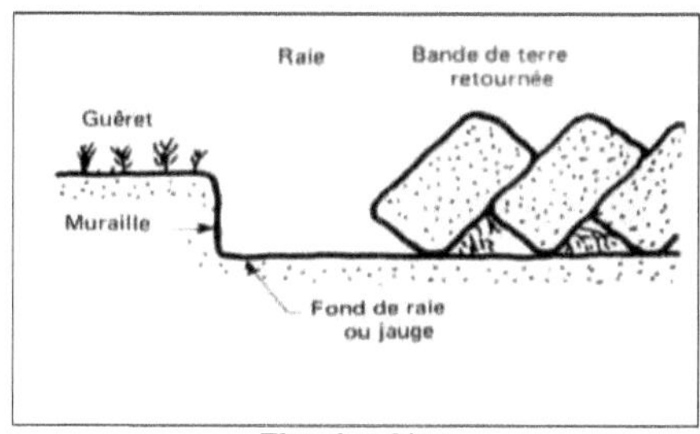

Flat ploughing

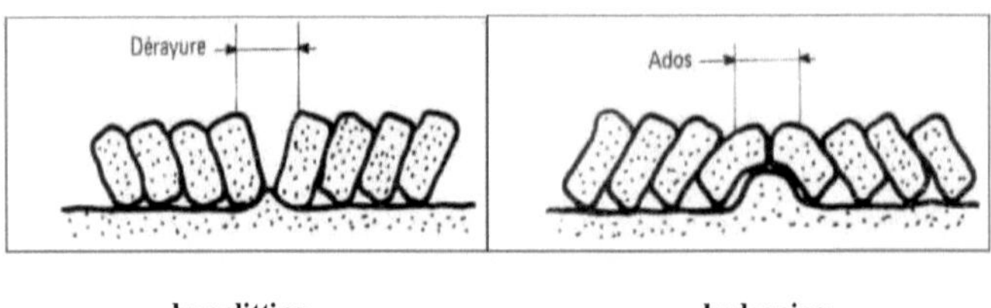

by splitting   **by leaning**

Fig 3. Ploughing techniques at plot level

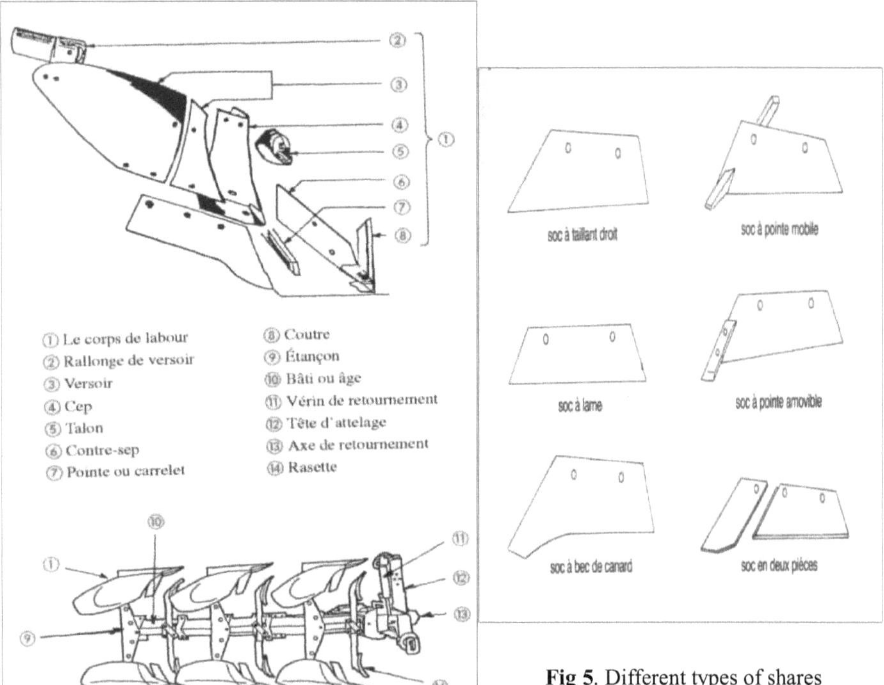

Fig 5. Different types of shares

Fig 4. Main parts of the plough

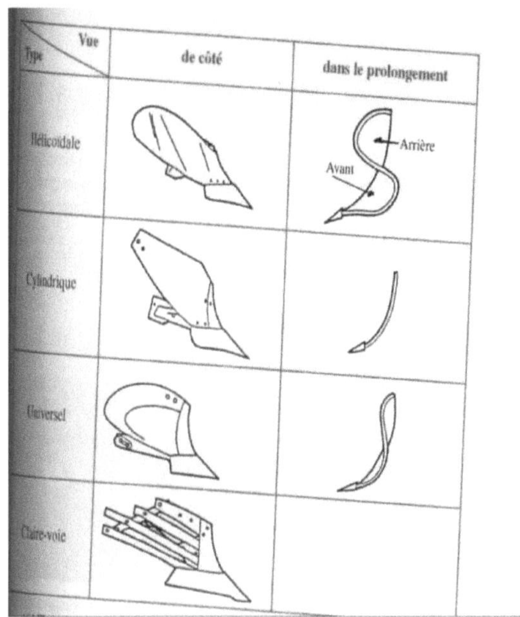

Fig 6. Different types of moldboards

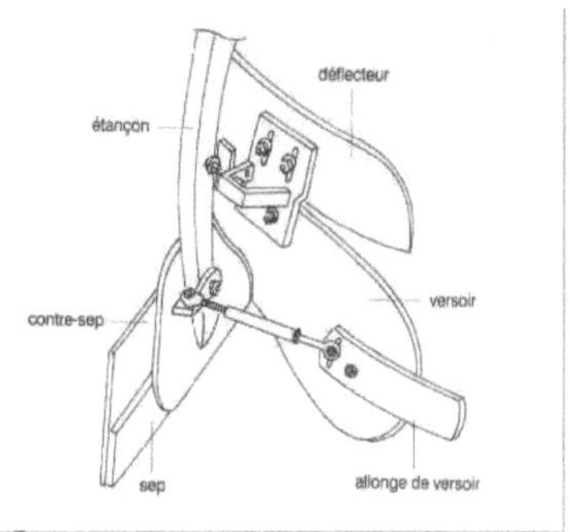

Fig 7. Additional equipment for moulders

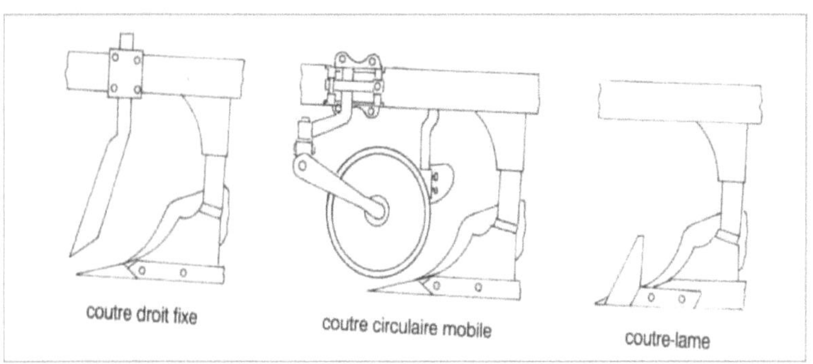

Fig 8. Different types of coulters

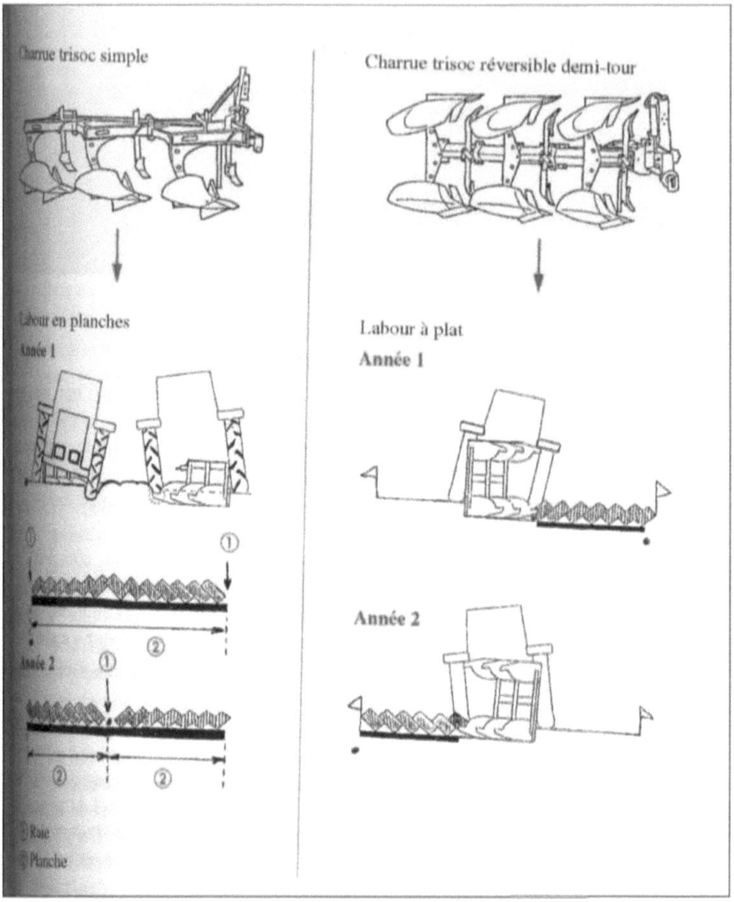

Fig 9. Classification of ploughs and types of ploughing

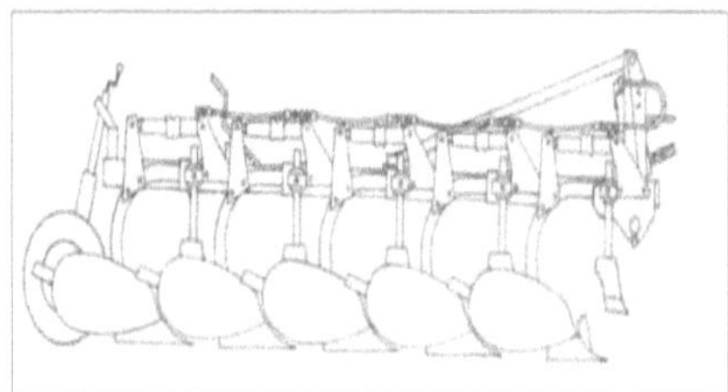

Fig 10. Simple plough with five bodies

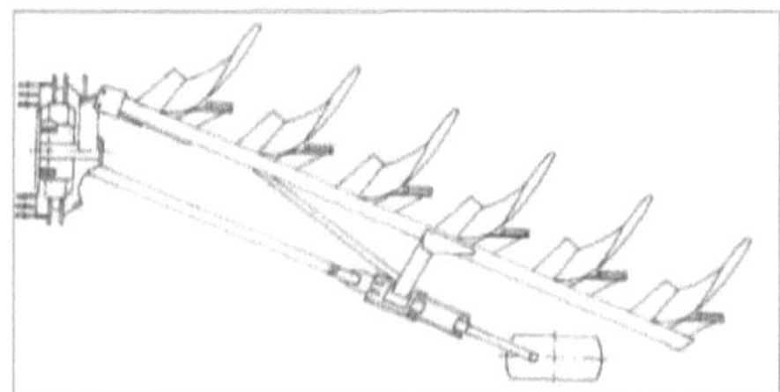

Fig 11. Semi-mounted plough with single support wheel

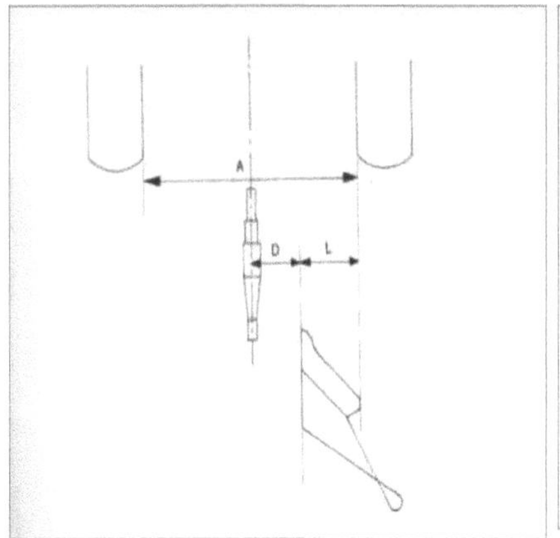

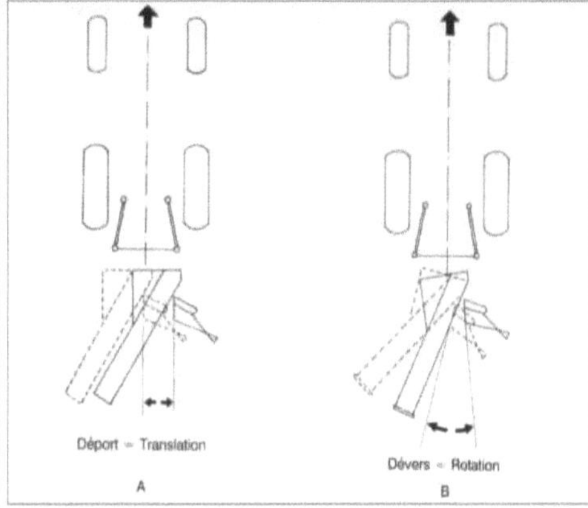

Fig 12. Determination of the tire gap

Fig 13. Adjustment of offset and tip tilt

L = Working width per body

D = Distance between the first counter-sep and the tractor axis

A = The tire gap

A = 2 (L + D)

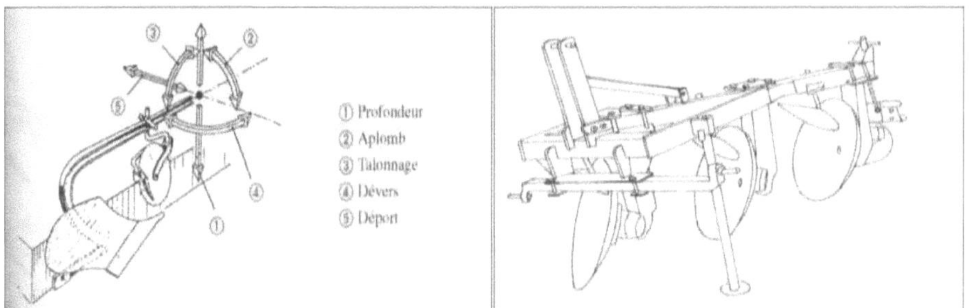

Fig 14. Plough settings

Fig 15. Single mounted plough with three discs

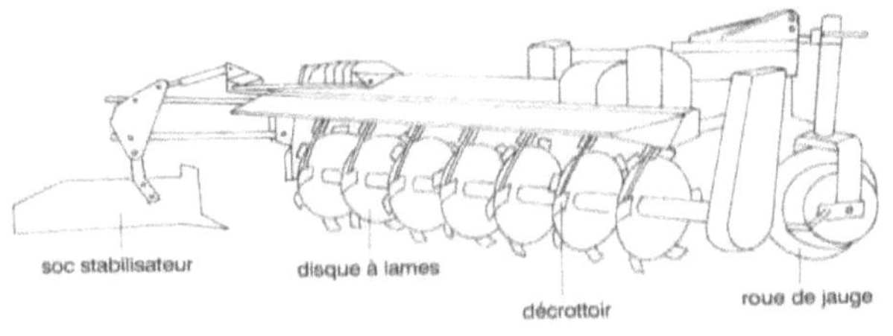

Fig 16. Controlled disc plough

Fig 17. Subsoiler

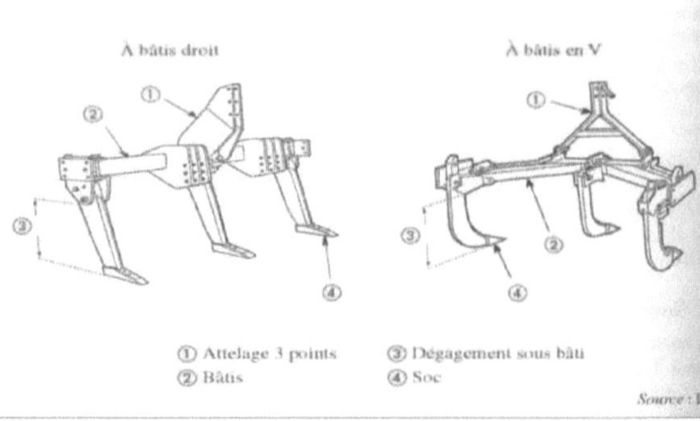

Fig 18. Decompactor

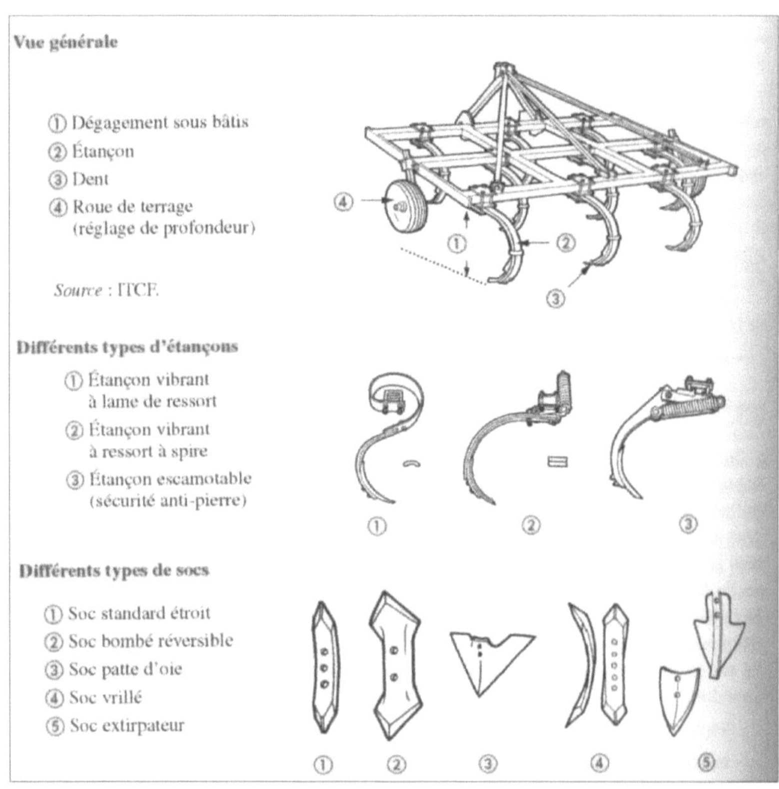

Fig 19. Chisel

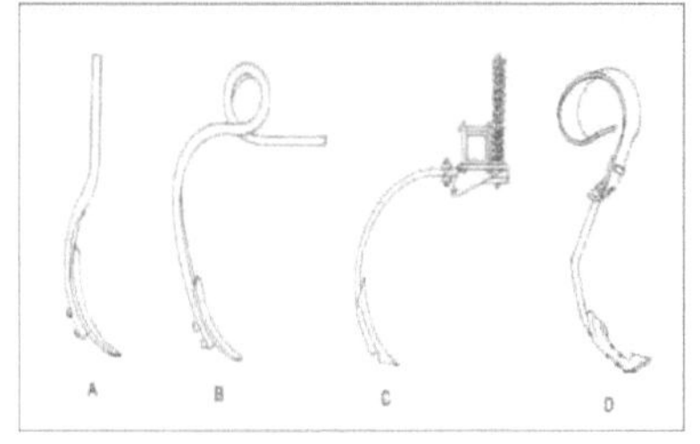

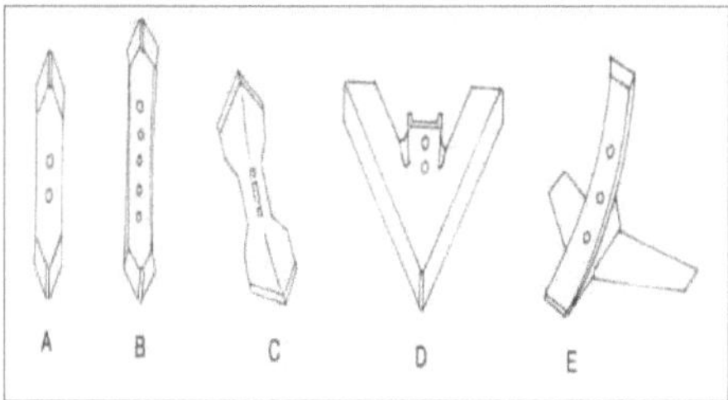

Fig 20. Different types of cultivator tines
A. Rigid tooth, B. semi-rigid tooth,
C. flexible spring toothA
D. flexible flat spring tineC
winged share

Fig 21. Different types of plowshares
. Standard coulter, B. twisted coulter,
. curved share, D. crow's foot share, E.

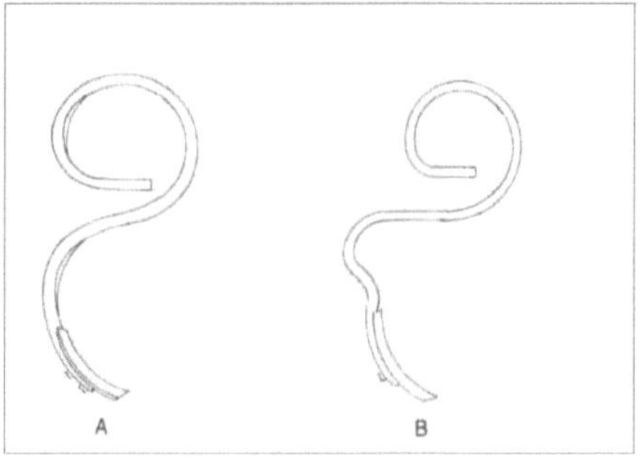

Fig 22. Vibro-cultivator teeth
A. Single curved tooth, B. Double curved tooth

Fig 23. Rolling spade cultivator

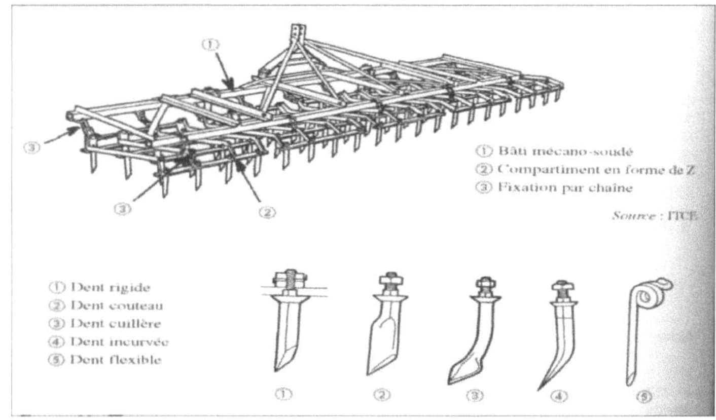

Fig 24. Harrow

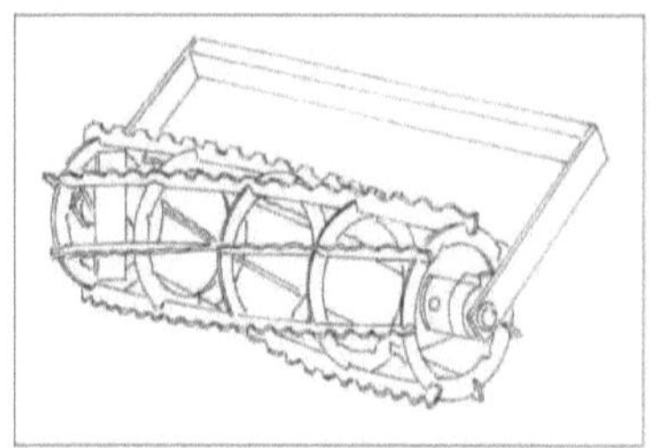

Fig 25. Rolling cage harrow

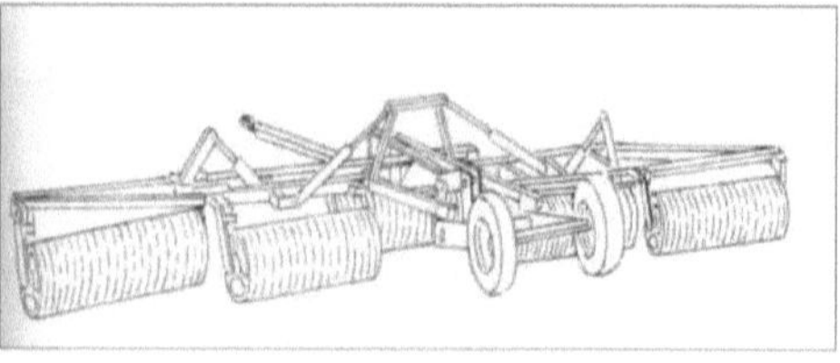

Fig 26. Roller assembly

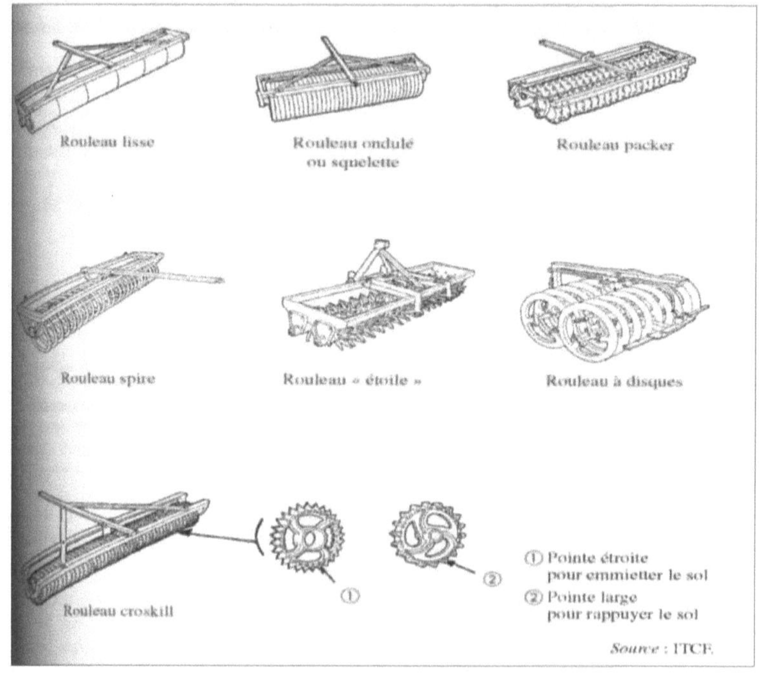

Fig 27. Main roller models

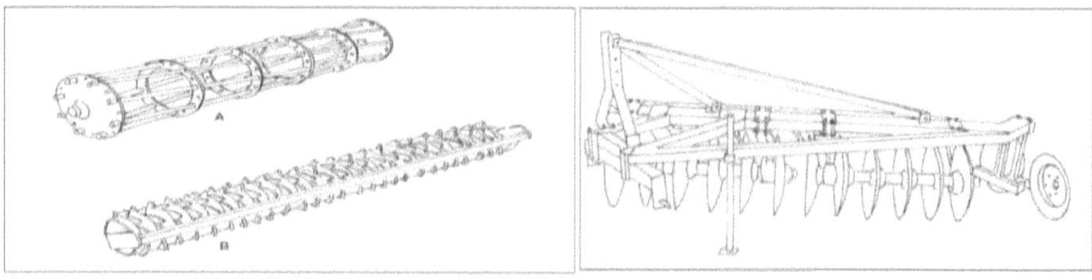

Fig 28. Rollers integrated into the controlled tools
A. Bar roller, B. Packer roller

Fig 29. Disc harrow

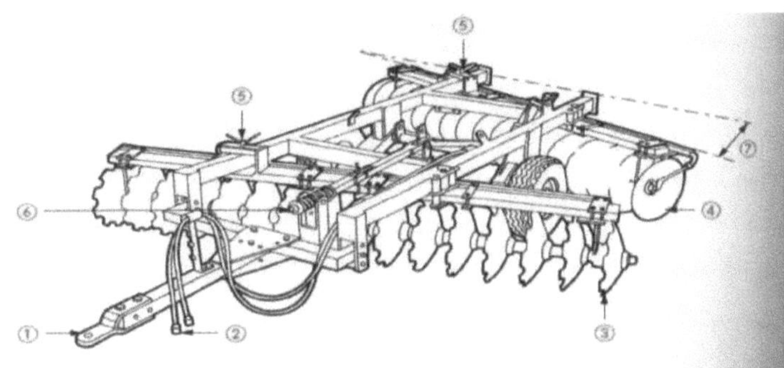

Fig 30. V" disc sprayer

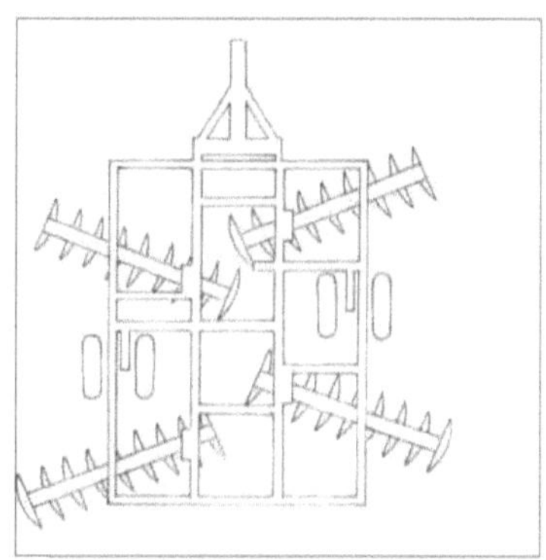

Fig 31. Tandem sprayer

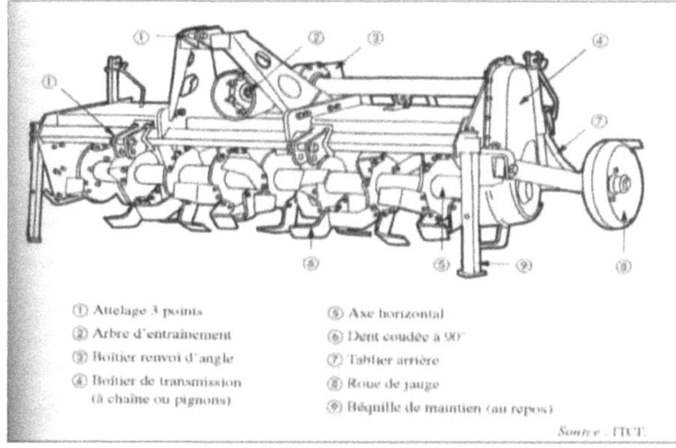

Fig 32. Rotary cultivator with horizontal axis

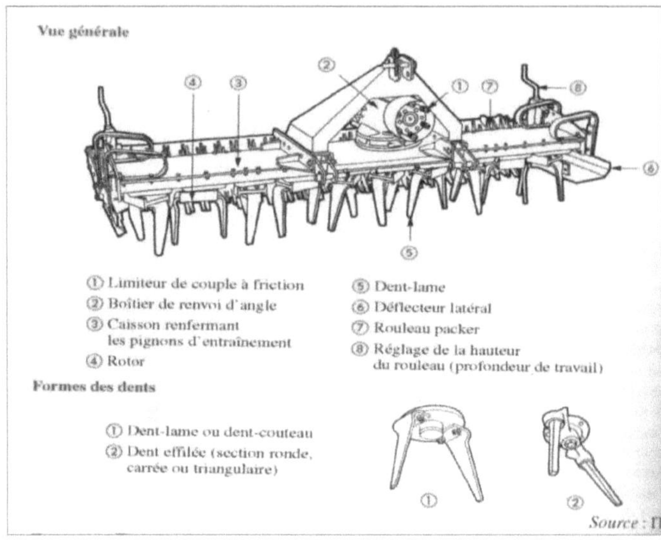

Fig 33. Rotary harrow

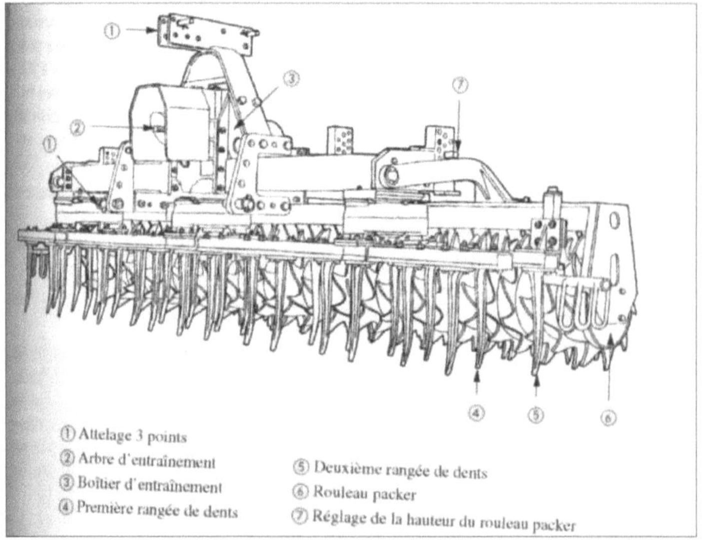

Fig 34. Reciprocating
harrow

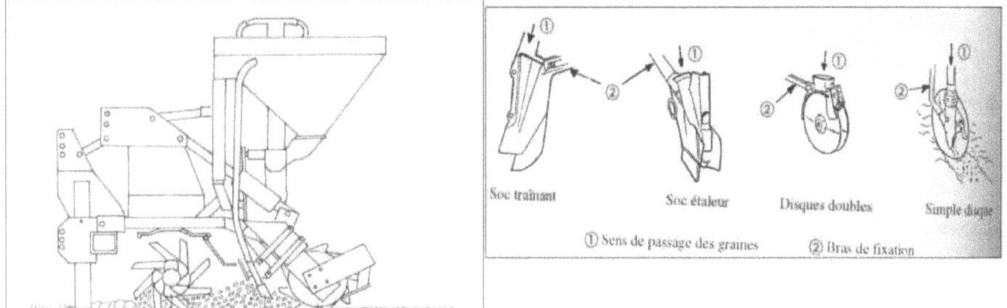

Fig 36. Burying devices

Fig 35.1.loosening tine, 2.tine rotor, 3.seed distribution, 4.packer roller

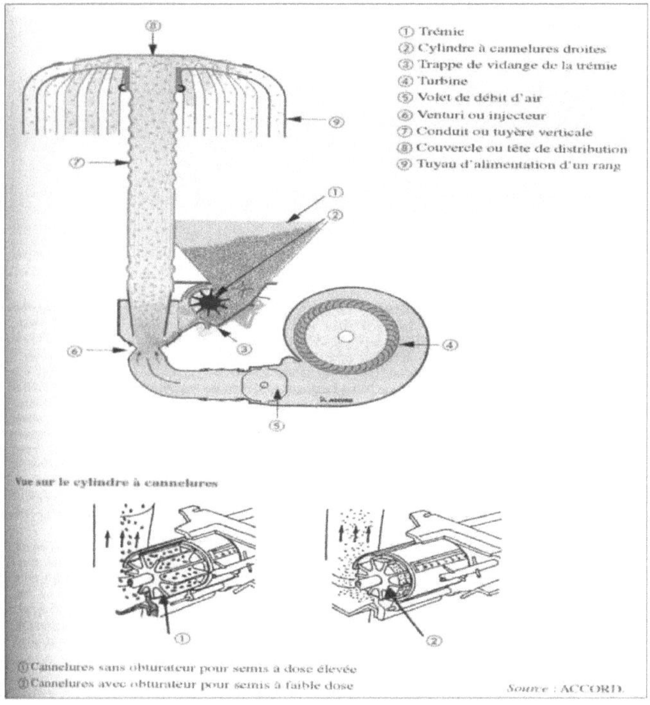

Fig 37. Pneumatic system of a seed drill with mechanical centralized distribution and pneumatic transport

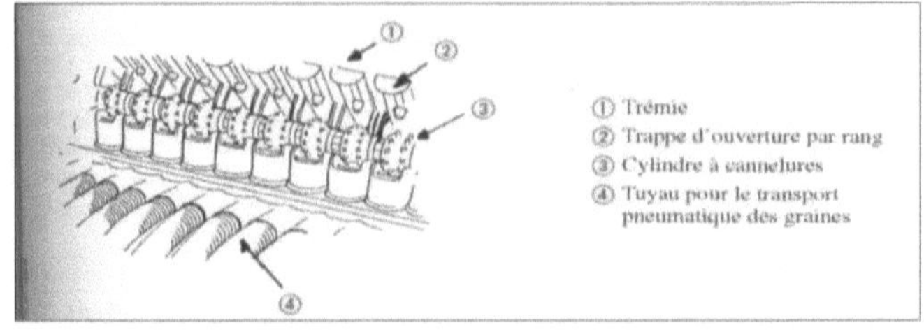

Fig 38. Principle of a pneumatic multidistribution

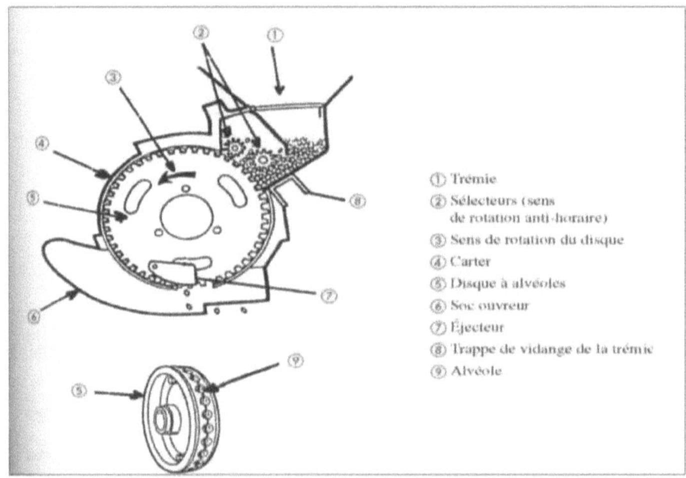

Fig 39. Principle of a mechanical distribution with a vertical disc with cells

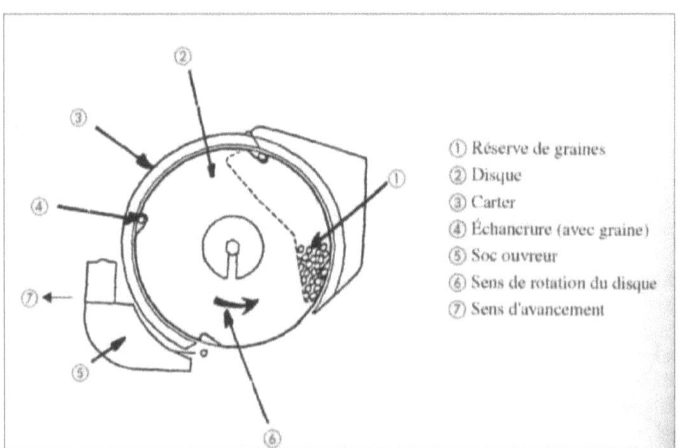

Fig 40. Principle of a mechanical distribution with vertical notched disc

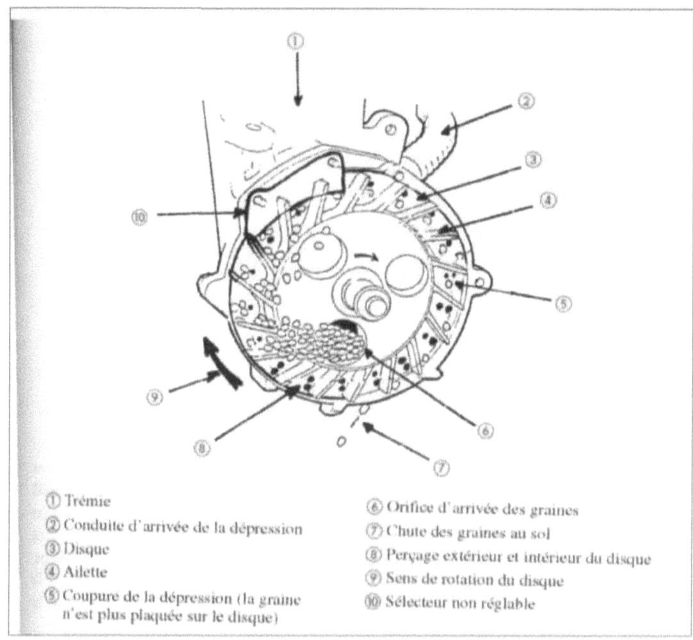

41. Principle of a pneumatic disc and vane wheel distribution

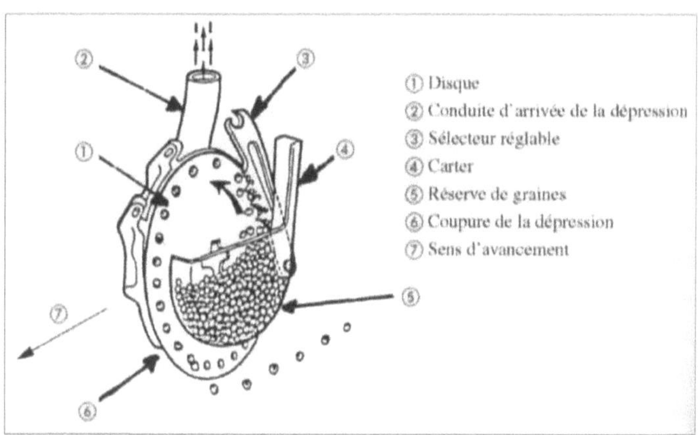

Fig 42. Principle of a pneumatic distribution with vacuum disc

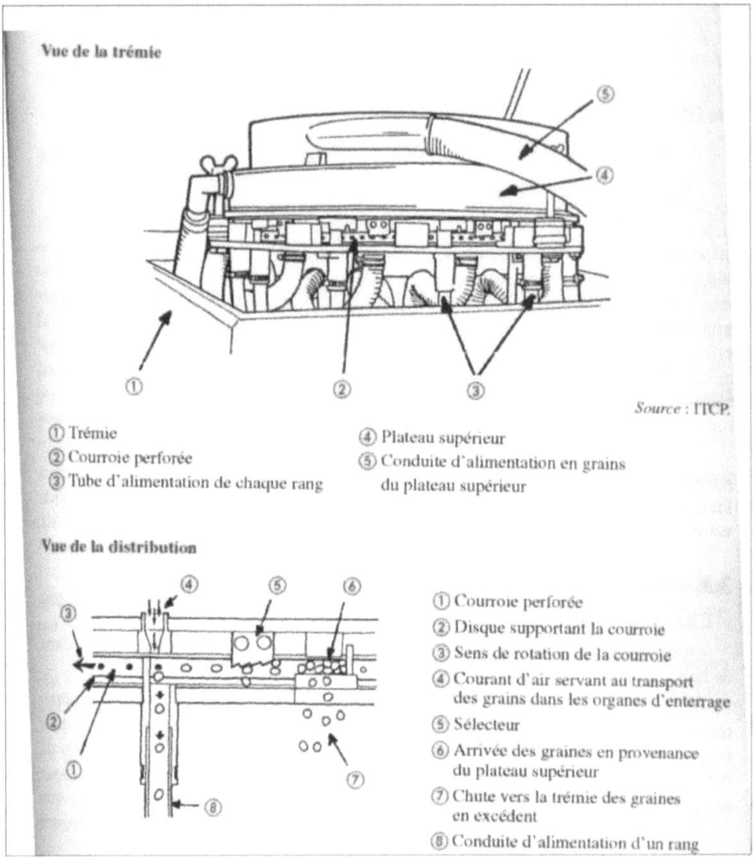

Fig 43. Multidistribution with vacuum

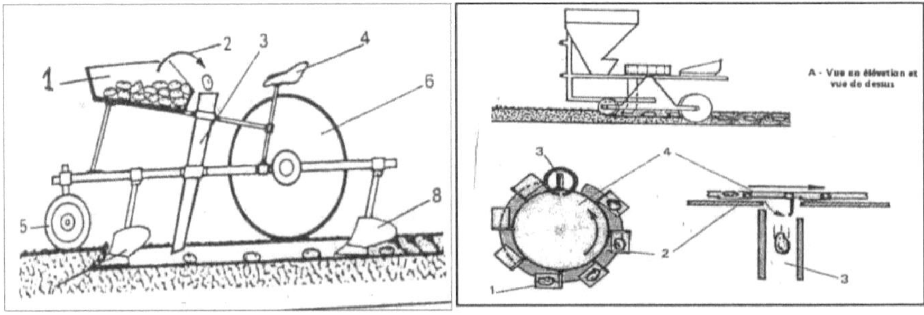

Fig 44   **Fig 45**

Fig 44. Principle of a tuber planter with entirely manual feeding

Fig 45. Tuber planter with manual feeding and automatic distribution by means of a horizontal rotary table

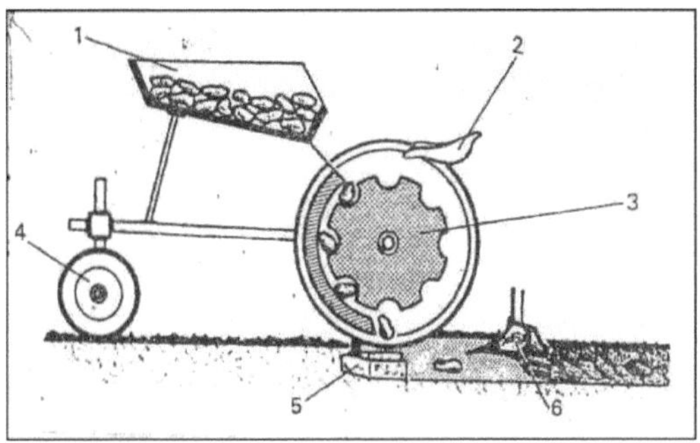

Fig 46. Tuber planter with AM and DA by vertical rotary tray

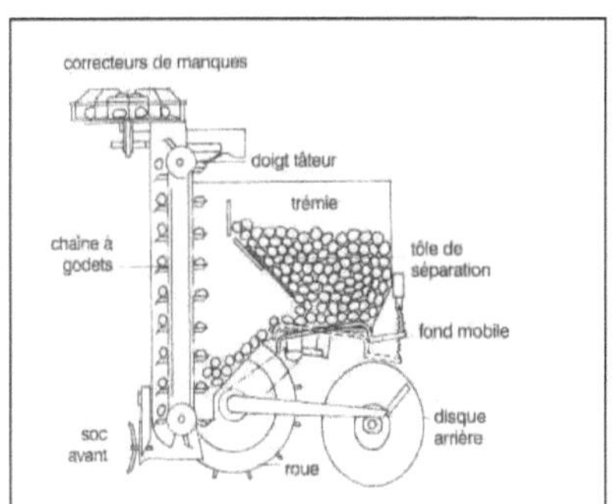

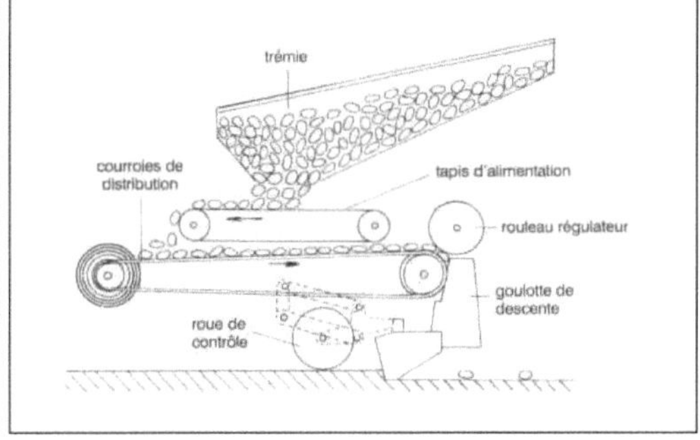

Fig 47. Chain planter with one row of buckets

Fig 48. Planter with belt drive

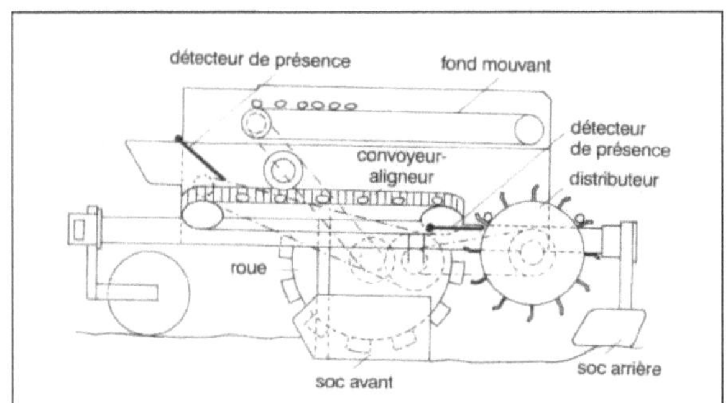

Fig 49. Feeding wheel planter fed by conveyor-aligner

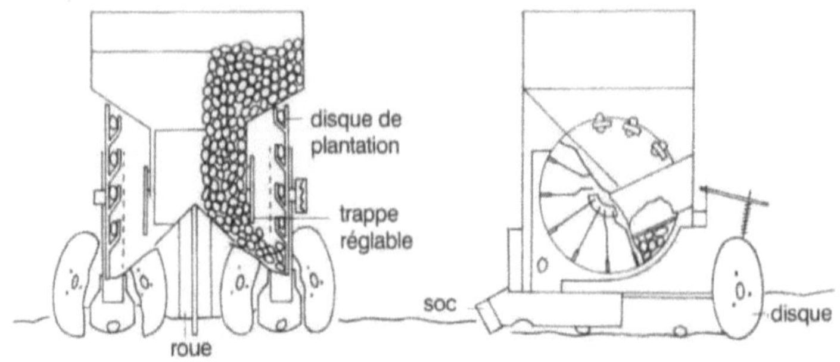

Fig 50. Planter with disc system and puller fingers

Study of crop maintenance tools

Fertilization and treatment equipment

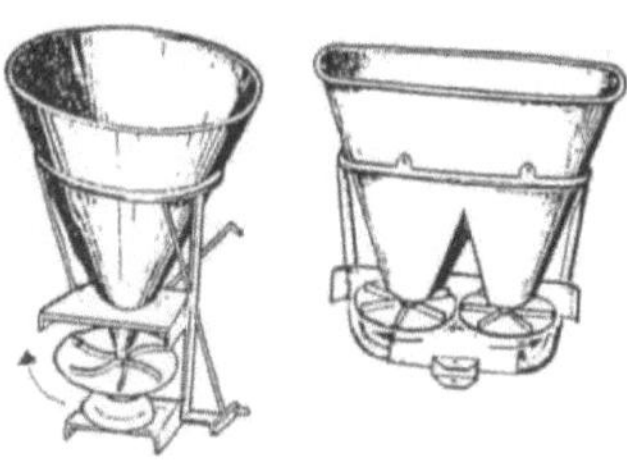

Fig 1. schematic view of a single and double disc centrifugal distributor

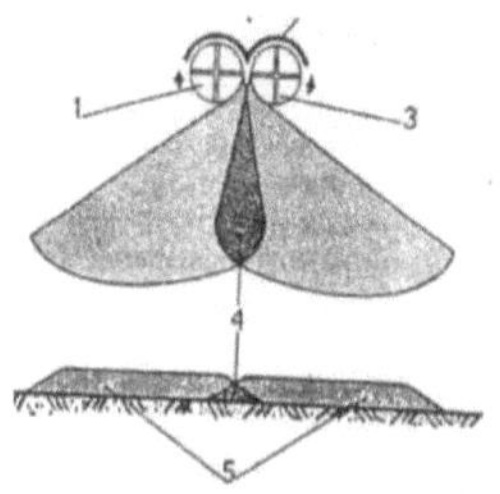

Fig 2: Fertilizer distribution by a double disc centrifuge

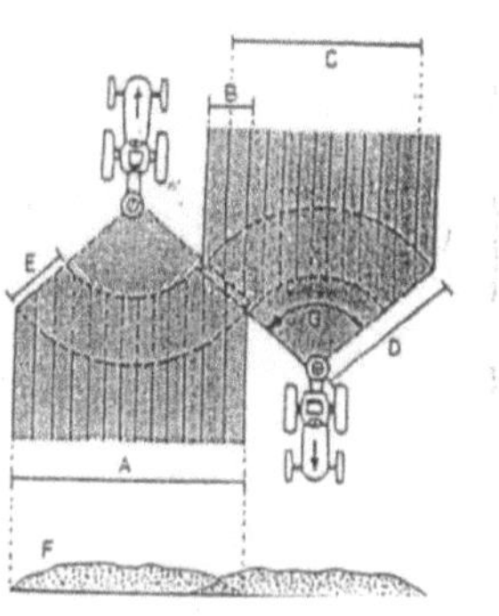

Fig 3. Spreading

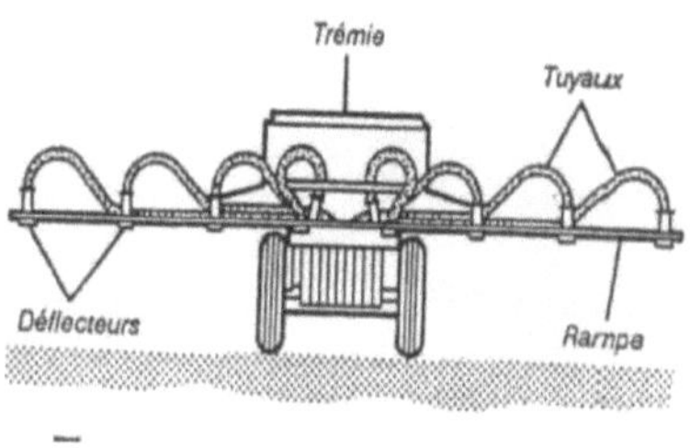

Fig 4. Rear view of a pneumatic distributor

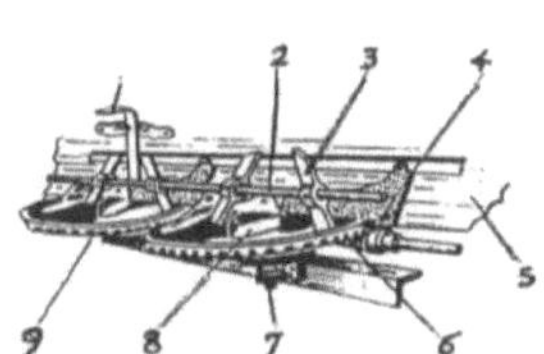

Fig 5. Plate fertilizer distribution

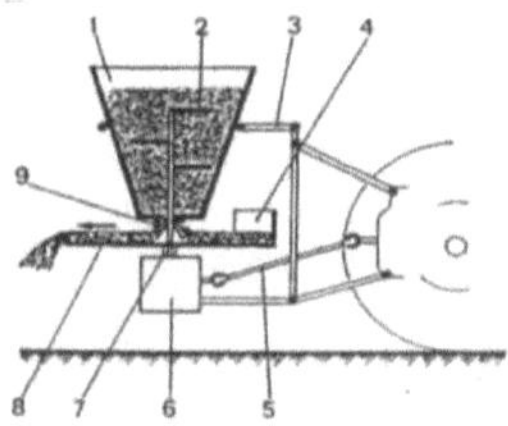

Fig 6. Single disc centrifugal fertilizer distribution

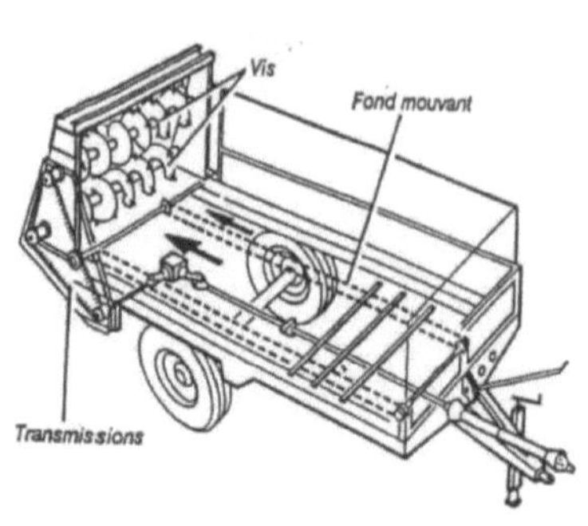

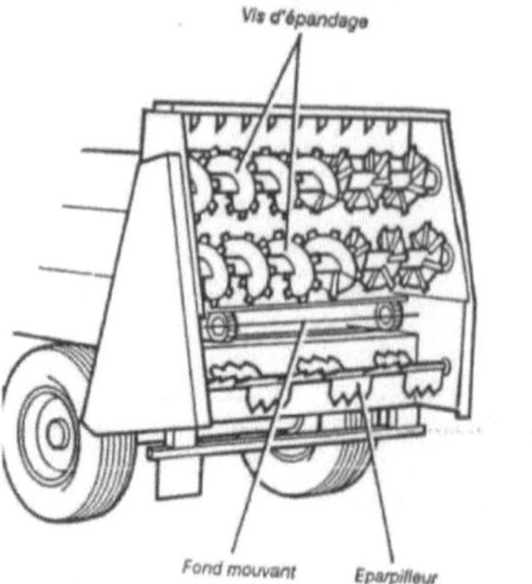

Fig 7. Manure spreader with rear spreading

Fig 8. Manure spreader with screw

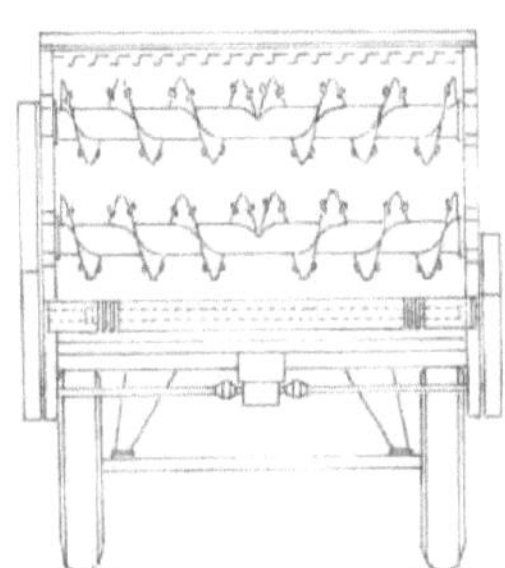

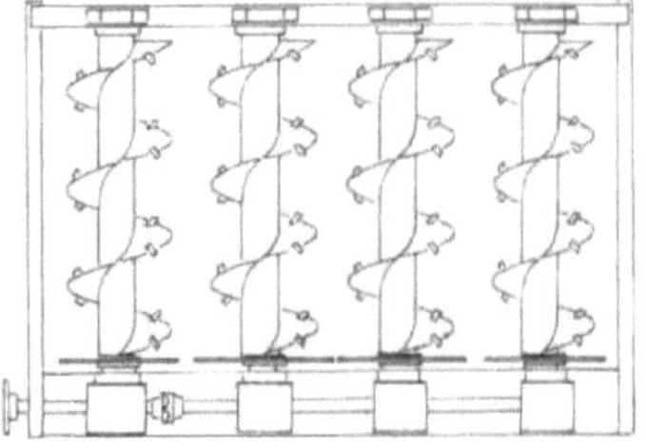

Fig 9. Backhoeing with horizontal reels

Fig 10. Rear spreading with vertical reels

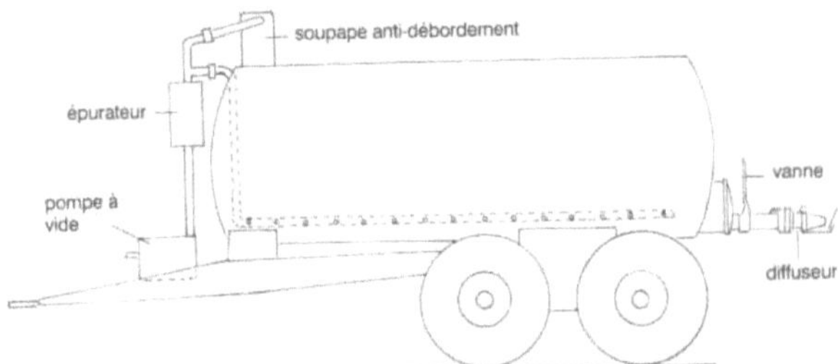

Fig11. Diagram of a lily spreader with vacuum pump

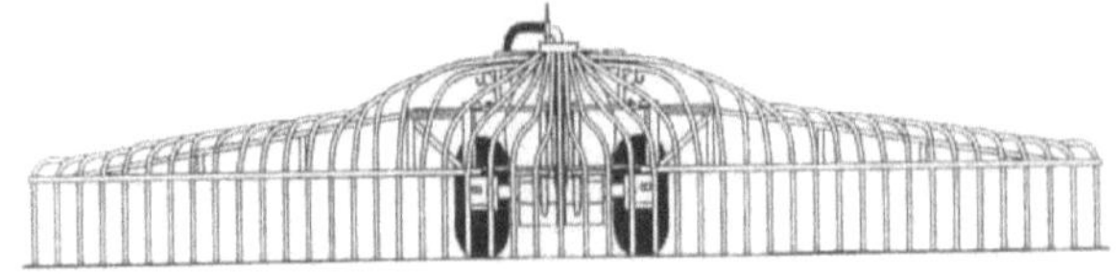

Fig12. Spreading ramp with trailing pipes

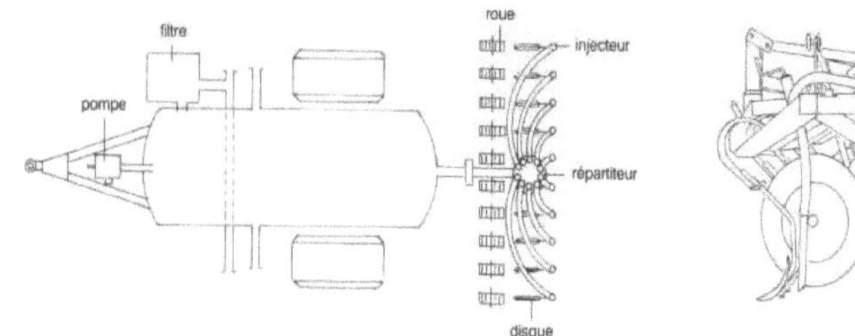

Fig 13. Manure burier for grassland

Fig 14. Spreader with a "crop" type manure burier

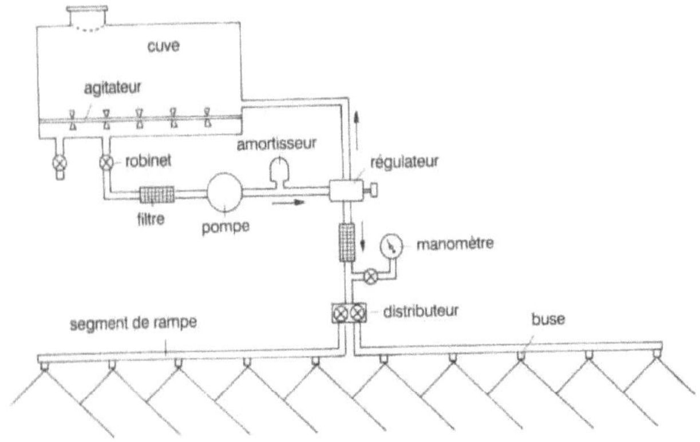

Fig 15. General diagram of a sprayer

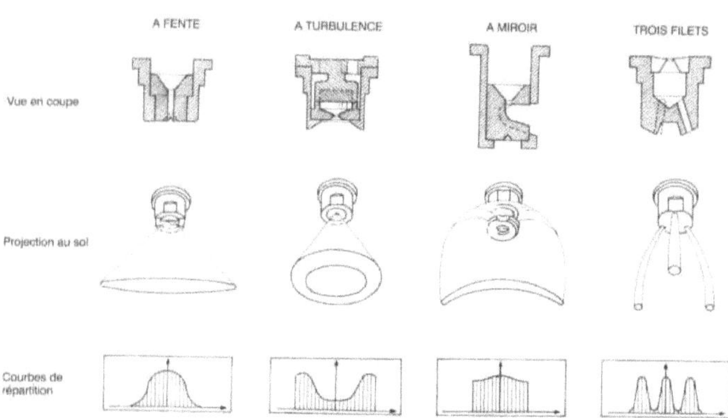

Fig 16. The different types of nozzles

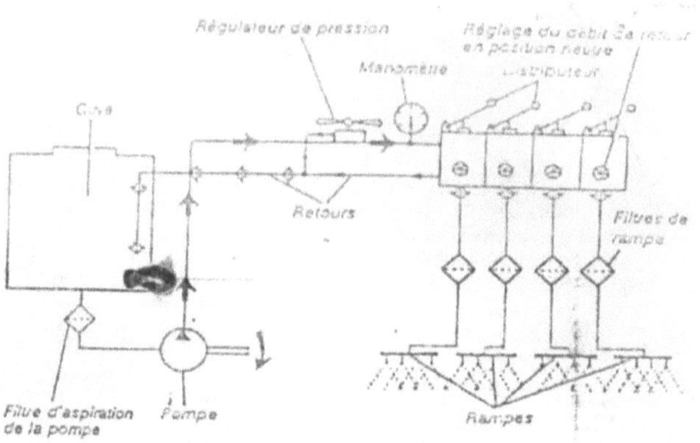

Fig 17. Diagram of the circuits of a pressure sprayer

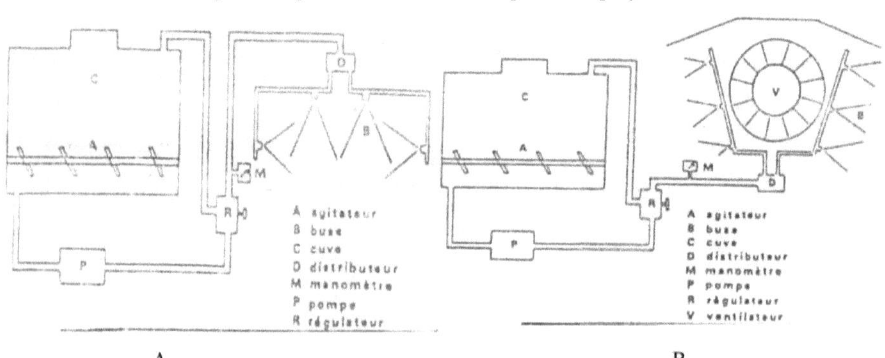

A

B

Fig 18. A. Pressure sprayers with spray pattern   **B.** Airborne

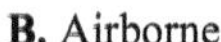

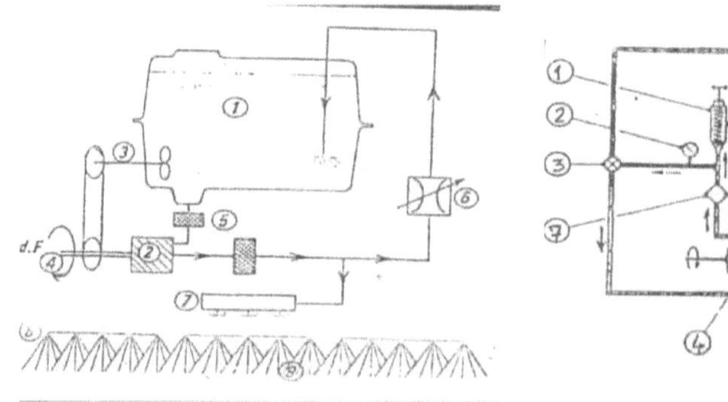

Fig 19. General diagram of a sprayer   **Fig 20.** Operation of a conventional liquid pressure sprayer

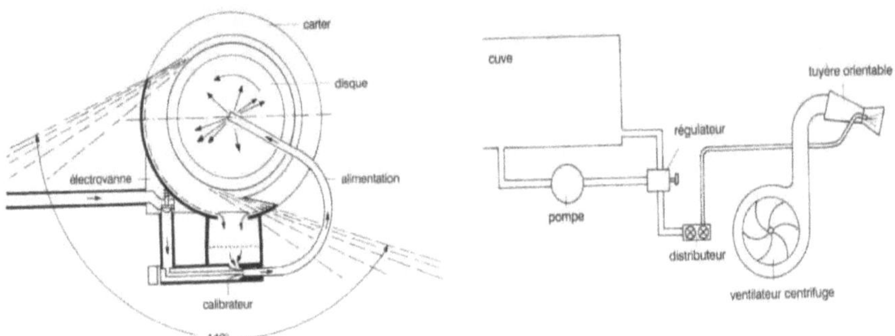

Fig 21. Vertical Axis Rotating Diffuser

Fig 22. Operating principle of a pneumatic sprayer

Mowing and harvesting equipment

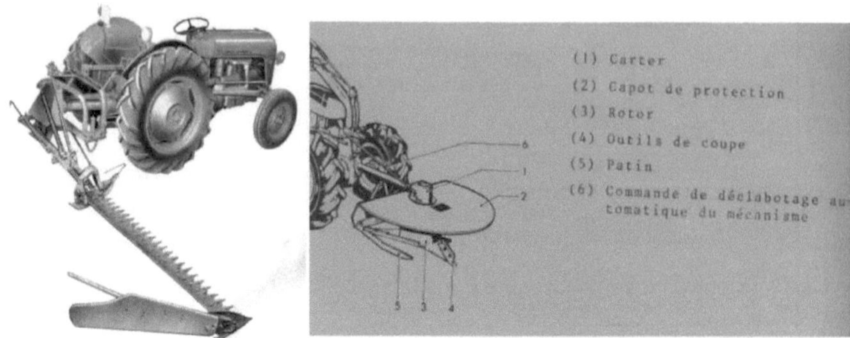

Fig 1: Reciprocating or bar mower

Fig 2: Presentation of a rotary mower

Fig 3. Rotary drum mower

Fig 4. Roller conditioner device

Fig 5. Finger rotor conditioner

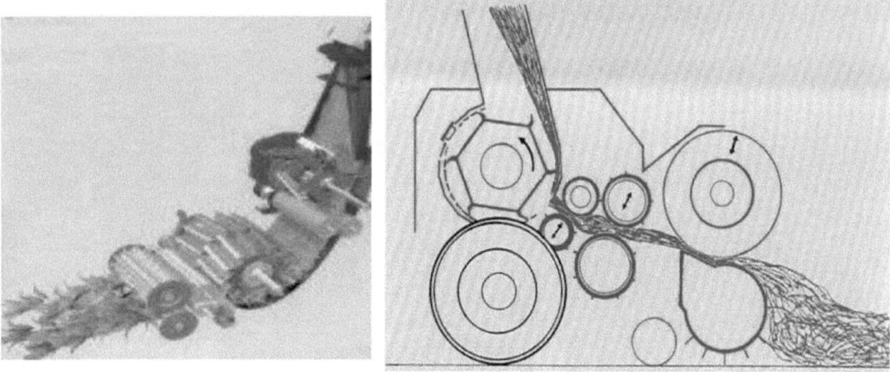

Fig 6. Operation of a thin-cutting forage harvester

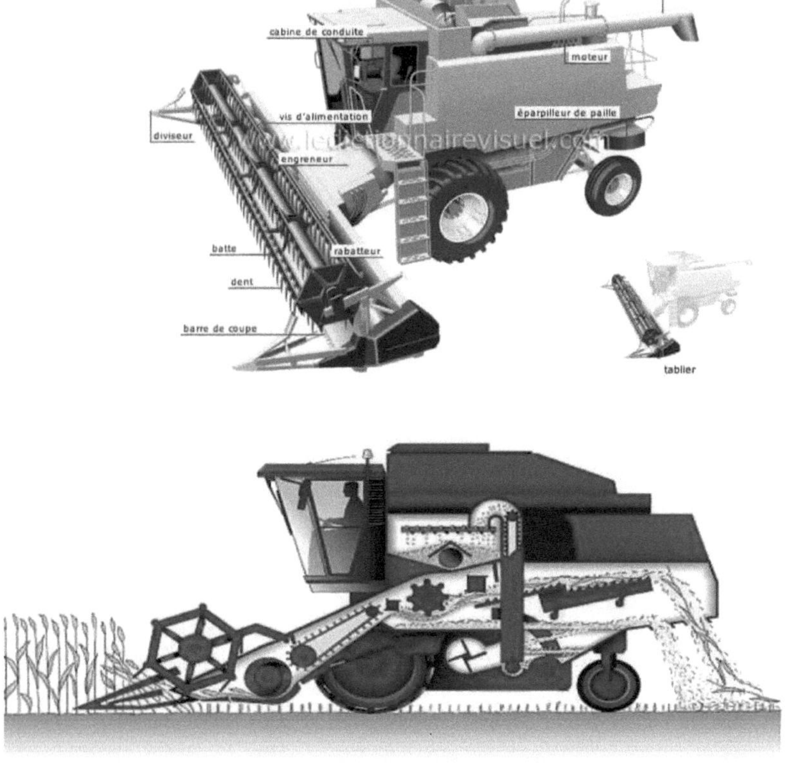

Fig 7,8. Combine harvester

Printed by Books on Demand GmbH, Norderstedt / Germany